MICROCOMPUTERS IN GEOMETRY

Mathematics and its Applications

Series Editor: G. M. BELL, Professor of Mathematics, King's College (KQC), University of London

Statistics and Operational Research

Editor: B. W. CONOLLY, Professor of Operational Research, Queen Mary College, University of London

Mathematics and its applications are now awe-inspiring in their scope, variety and depth. Not only is there rapid growth in pure mathematics and its applications to the traditional fields of the physical sciences, engineering and statistics, but new fields of application are emerging in biology, ecology and social organisation. The user of mathematics must assimilate subtle new techniques and also learn to handle the great power of the computer efficiently and economically.

The need of clear, concise and authoritative texts is thus greater than ever and our series will endeavour to supply this need. It aims to be comprehensive and yet flexible. Works surveying recent research will introduce new areas and up-to-date mathematical methods. Undergraduate texts on established topics will stimulate student interest by including applications relevant at the present day. The series will also include selected volumes of lecture notes which will enable certain important topics to be presented earlier than would otherwise be possible.

In all these ways it is hoped to render a valuable service to those who learn, teach, develop and use mathematics.

For complete series list see back of book

MICROCOMPUTERS IN GEOMETRY

ADRIAN OLDKNOW, B.A., M.A., M.Tech., F.I.M.A., F.B.C.S.
Principal Lecturer in Mathematics and Computing
West Sussex Institute of Higher Education
Bognor Regis

ELLIS HORWOOD LIMITED
Publishers · Chichester

Halsted Press: a division of
JOHN WILEY & SONS
New York · Chichester · Brisbane · Toronto

First published in 1987 by
ELLIS HORWOOD LIMITED
Market Cross House, Cooper Street,
Chichester, West Sussex, PO19 1EB, England
The publisher's colophon is reproduced from James Gillison's drawing of the ancient Market Cross, Chichester.

Distributors:
Australia and New Zealand:
JACARANDA WILEY LIMITED
GPO Box 859, Brisbane, Queensland 4001, Australia
Canada:
JOHN WILEY & SONS CANADA LIMITED
22 Worcester Road, Rexdale, Ontario, Canada
Europe and Africa:
JOHN WILEY & SONS LIMITED
Baffins Lane, Chichester, West Sussex, England
North and South America and the rest of the world:
Halsted Press: a division of
JOHN WILEY & SONS
605 Third Avenue, New York, NY 10158, USA

British Library Cataloguing in Publication Data
Oldknow, A.J.
Microcomputers in geometry. —
(Ellis Horwood series in mathematics and its applications)
1. Geometry — Data processing
2. Microcomputers
I. Title
516'.0028'5416 QA447

Library of Congress Card No. 86–27874

ISBN 0–85312–877–4 (Ellis Horwood Limited — Library Edn.)
ISBN 0–7458–0185–4 (Ellis Horwood Limited — Student Edn.)
ISBN 0–470–20805–8 (Halsted Press — Library Edn.)
ISBN 0–470–20814–7 (Halsted Press — Paperback Edn.)

Phototypeset in Times by Ellis Horwood Limited
Printed in Great Britain by Unwin Bros. of Woking

Table of contents

Preface

In the earlier book, *Learning Mathematics with Micros*, just one chapter was given over to computer geometry. This new book is based upon a variety of ideas in two-, three-, and even four-dimensional geometry that have been explored with teachers, students and pupils, since the publication of that earlier book in 1983. In this time domestic microcomputers have proliferated, and their prices have fallen while the facilities they offer have grown. Now there are many inexpensive microcomputers in High Street stores offering graphical facilities far superior to the biggest computers of twenty years ago.

This book is not intended to be comprehensive or authoritative. It is aimed to give the reader a feel, through many "worked examples", of how a wide variety of geometric ideas can be explored through computer programs. I hope it conveys something of the fun I have had myself in discovering and interpreting bits of geometry.

Thus it is hoped that readers with an interest in geometry but little experience or confidence in computer programming will feel encouraged to use the micro as an exploratory yool.

Similarly it is hoped that readers with skills in computing but little experience in geometry (and related mathematics) will find that there is considerable pleasure and fun to be had with exploring some powerful mathematical ideas.

There has been a welter of marvellous invention in geometry over the centuries since the Greeks and much of this is accessible to us in books. With a micro (alongside other cheaper and useful artefacts such as pencils, rulers, compasses, erasers, etc.) we can gain access to the worlds of these mathematicians and make the ideas "come alive". If we are lucky we might also discover some results or shapes that are quite new.

However, the mathematical level of this book seldom strays beyond the kind of skills usually associated with A-level mathematics in Britain and the ideas have all been tried out with15–18 year olds. Although many of these ideas (curves in space, surfaces, hypercubes, etc.) are often not "on the syllabuses" the same is not true of the underlying techniques of coordinate geometry, vector algebra, matrix transformations, etc. Thus I trust the book will appeal to students of mathematics in sixth forms and colleges who have access to a microcomputer.

Modern microcomputers differ considerably in their commands for handling graphics but the vast majority include a few simple commands to clear the screen, plot a point, draw a line and, perhaps, shift the origin. The book includes large numbers of example programs that can be easily converted to run on any such micro.

Programs developed in the text are in a simple form of BBC Basic which will run altered on the BBC, Master ane Electron micros. The programs are carefully designed to enable the easiest possible conversion to run on most common micros, including the IBM PC, Apple Macintosh, Apple II, Sinclair ZX-Spectrum, Research Machines 380/480Z and RM Nimbus.

My idea of a super way of passing a wet weekend is to tuck myself away with my favourite micro, a good geometry book (such as Lockwood's *A Book of Curves*), an endless supply of coffee and then to just "doodle" with some programs. I hope that after reading this book you will add this recreation to your list of hobbies. It is far less frustrating than golf, but be warned — it is even more antisocial !

ACKNOWLEDGEMENTS

I owe my interest in geometry to Cyril Nobbs (at the City of London School). I am indebted to the undergraduate mathematics course at the University of Oxford of the mid-1960s for (a) having baffled me completely about geometry and (b) having completely ignored computers. I owe my interest in Computer Graphics to Brunel University, where I first learned about it, and to the Computer Aided Design Centre in Cambridge, where I was allowed to try to apply it. I owe my interest in trying to communicate geometrical ideas to the inspiration of Trevor Fletcher.

Over the past few years I have been able to work with teachers, students and pupils using microcomputers and I am grateful to the many people without whose support this would not have been possible. These include: the Department of Industry (now the DTI), Research Machines of Oxford, Acorn Computers of Cambridge, and the Microelectronics Education Programme (MEP) of the Department of Education and Science. I am indebted to the many teachers, students and pupils who have collaborated in this work (too many to name here), but most especially to John White and Professor David Johnson who have contributed so much.

With regard to the book I am grateful to Ellis Horwood (a) for having persuaded met to write in the first place, (b) for allowing me to do it again and (c) for maintaining to publish such a wide range of exotic literature. The manuscript was typed on a BBC micro using the Computer Concepts

"Wordwise-Plus" system and most of the illustrations were produced on a Linear Graphics "Plotmate" plotter.

Chichester, 1986 Adrian Oldknow

To Cyril, Trevor & John

List of programs

Introduction

Geometry is in rather a strange state at the moment. There is an enormous wealth of "pure" geometry embracing constructions. Euclid, coordinate geometry, vectors, projective geometry, transformations, differential geometry and non-Euclidean geometries, to mention just a few. It is an area of mathematics that has fascinated many great mathematicians and yet its prominence in the mathematics syllabuses of schools and colleges has declined. On the other hand, since the advent of commercial computers, a new "applied" subject of Computational Geometry has emerged to underpin computer-aided design (CAD) which, at present, seems to be restricted to a few postgraduate courses and a small body of working professionals. Ironically the mathematical foundations of many techniques from Computational Geometry are not particularly difficult, mathematically, and are well within the reach of many students.

Now that the domestic market for microcomputers has been well established many people interested in mathematics have at their disposal a tool of phenomenal power that would have been unbelievable just a few years ago. Unfortunately the impression is that this resource is extremely under-used. There appear to be two quite distinct categories of computer owners: the programmers and the non-programmers.

The image of the programmers is of an impenetrable group who know their particular micros inside out, talk "hex", read strange magazines, hack into places they should not go and hanker after making a fortune by writing the best computer game ever. The non-programmers admit to being baffled by things technical and prefer to treat the micro like a record player or video recorder — buying (or acquiring) software which may be largely recreational (Invaders, Frogger, Elite and the like) or functional (word-processing, data-bases, accounts, etc.).

At the heart of this apparent divide is the place of programming. Many people who have studied mathematics have encountered, during their training, a short course on some programming language (usually Fortran) which has served to convince them that programming is impossible, irrelevant and actually harmful. This common experience takes a lot of undoing. Programming is now beginning to achieve acceptance as a recognised component in mathematical education and it has much ground to make up. Unless learning to program can be approached in a more constructive and effective way than has usually been the case it is unlikely to make much impact.

Yet programs are merely bits of writing in some more or less formal language and should be within the reach of most who would regard themselves as mathematical. Like other forms of writing in "foreign" languages there is a great difference in the skills required to read, or interpret, and those needed to write, or generate. Most teachers of languages would expect to give their students a good grounding in reading and interpreting before expecting much by way of writing and generating.

Like all analogies this one probably does not bear too detailed a pursuit. The key element in learning about programming is the immediate feedback through interaction with a micro. You could just take a small program from a book (such as this one), type it in to a micro (if it was not supplied on disc or cassette), run it and sit in amazement pondering the wisdom and skill of its author. On the other hand you could study the anatomy of the program by examining the effect that small changes to it produce. If the program is mathematical you might also begin to wonder about some mathematical properties of the output.

It is in that spirit that this book is offered. Some editorial decisions have had to be made. The most obvious ones concern the programming language(s) and the particular microcomputer implementation(s) to be used. The language issue is the more straightforward. Without going into technicalities the principle of easy interaction means that the language has to be one which is interpreted (like Basic or Logo) rather than one which is compiled (like Fortran and most versions of Pascal). The language needs to be one which is accessible to most readers with their own micros, and that usually means the one given away "free' with the micro. At the moment most domestic microcomputer manufacturers supply a version of Basic "built-in" to the machine in the form of a ROM plugged inside the case. With the newer generations of 16-bit micros it may become more commonplace to buy your micro with, say, two or three languages (perhaps Basic, Logo and Pascal) supplied on disc or cartridge together with the other "bundled software" that is now commonly part of the package deal.

The choice of micro is more complex. In principle, programs written in Basic should be able to run on all micros with Basic interpreters, but, of course, the practice is far from the case. Firstly there is no "standard" Basic at present other, perhaps, than the original formulation of Kemeny and Kurtz of more than twenty years ago. As manufacturers have added extra facilities to their micros (graphics, sound, electronic gadgets, etc.) they have "tacked on" extra commands to control them. In particular most micros now offer some form of "high resolution" graphical display in colour. Needless to say they all have their own highly individual ways of using them.

I have taken the decision, then, to write the programs used in this book in Basic. The programs are written in a style designed to make them as readable as possible, easily convertible for other micros and to other languages, with an eye on their computational efficiency. For this purpose the BBC Basic language was the obvious choice as it now seems to be setting the standard for more recent versions of Basic. However, to make the programs usable by a wider class of readers than just BBC micro owners, the programs use a restricted range of commands that can be easily converted into equivalents for other popular micros such as the Apple Mac, the IBM PC, the Sinclair ZX-Spectrum, the Apple II and Research Machines 380/480Z and RM Nimbus.

Chapter 1 sets out to compare the simplest graphical facilities of a number of common micros and to develop an approach to writing simple programs. It assumes no previous knowledge of programming and introduces the few simple, yet powerful, tools that are applied throughout the book (loops, conditionals, procedures, etc.). Most of this could be skipped altogether by those familiar with programming. It concludes with an important section on how programs throughout the book can be converted for other microcomputers — this should **not** be skipped.

Chapter 2 starts with a simple and well known example: the positions of a ladder slipping against a vertical wall envelope a curve — the Astroid. It takes this theme and develops a number of variations to do with envelopes and loci which provide algorithms for drawing circles, ellipses and polygons. These are applied to a number of interesting problems. it also introduces the idea of parametric methods for descriptions of curves and lines.

Chapter 3 explores a number of conventional representations of curves in parametric, cartesian, and polar forms. Armed with these techniques and the micro's "in-built" functions such as SIN, COS, LOG, EXP, you can invent all manner of new and exciting variations. It also shows how the techniques can be applied to equations of motion which use time as the parameter.

Chapter 4 is a collection of ideas for extending the programming techniques to a variety of interesting geometrical situations. It starts with some examples to do with conics and develops techniques for handling tangents, normals, curvature, arc-length and area. One section is devoted to using circles for generating limacons, epi- and hypo-cycloids, "spirographs" and inversions. A major section is devoted to the properties of a triangle and a comprehensive set of procedures are developed for finding key points, such as the incentre, circumcentre and orthocentre; drawing key lines, such as the angle bisectors, medians and altitudes; and describing key circles, such as incircle, circumcircle and 9-point circle.

Chapter 5 is mostly concerned with transformations and matrices. It looks at a number of interesting patterns that can be developed by rotations and reflections, and goes on to use shearing to demonstrate a well known result. A major section is devoted to developing a set of procedures to form a transformation geometry utility. The chapter concludes with examples of repeated matrix multiplication in finding eigenvectors and eigenvalues and in new ways of generating curves.

Chapter 6 introduces ideas from computational geometry in the form of Bezier curves, B-splines, natural splines and Hermite polynomials. The section includes techniques required to show a curve in space in orthographic, oblique and isometric projections.

Chapter 7 generalizes the ideas met for representing curves formed from functions of a single variable to surfaces formed from functions of two variables. Simple representations are developed for direct and parametric cartesian forms and are used to explore curves drawn on surfaces. A major section is then devoted to developing a set of procedures which build up to a sophisticated surface display and computation utility.

Chapter 8 is concerned with the representation of polyhedra. It starts with a simple skeletal representation and introduces techniques for transformations (such as stretches and rotations) and projections (such as isometric and perspective). Ways of introducing colour are explored, including the generation of 3D stereographic images. The representation is extended to include colouring the faces and to ways of detecting hidden lines. Data for the generation of the regular polyhedra (tetrahedron, cube, octahedron, icosahedron and dodecahedron) is generated. The chapter concludes with a section on extending the representation to 4D by illustrating transformations on a hypercube.

1

Some Basics

1.1 THE SCREEN

Most modern microcomputers have the capacity to output graphics. However, the commands for controlling this vary greatly from machine to machine and between computer languages. Fortunately, though, they all share a common "graphical output peripheral" — the domestic TV set (or a monitor that uses a similar principle). Thus the common ground between microcomputers is a "screen" consisting of a rectangle divided up into little drawing blocks, called **pixels.** Depending upon the sophistication of the micro there may be as many as 640 or as few as 150 pixel divisions across the screen and as many as 256 and as few as 100 up the screen. Each pixel may be "coloured" in any one of N colours selected from a "palette" of M colours, where, again, the values of N and M depend upon the sophistication of the micro.

The BBC computer has a number of graphics modes. The one we shall use mainly is called MODE 1 which provides a screen notionally divided into 320×256 'pixels' each of which can be in one of four colours selected from a palette of eight colours. This is a usual compromise: for a fixed size of screen the more pixels the finer the drawings ("higher resolution") but also the more computer memory required and hence the fewer the colours.

1.2 SCREEN GEOMETRY

In mathematics there are a number of systems for representing points and line-segments in a plane. For example, there is a Cartesian representation in which we fix two principal directions (**axes**) at right-angles to each other and define a scale of measurement along each. Points P are fixed by giving **coordinate** pairs (Px, Py) of measurements along the axes from the **origin** (0, 0). Line-segments PQ are defined by giving the coordinates (Px, Py), (Qx, Qy) of the points P and Q. This system is the one adopted for the graphics commands of the majority of micros.

Line-segments could equally as well have been defined by the coordinates (Px, Py) of a point on the line and a **displacement vector** (dx, dy). Conversions between the two systems is easily performed using the equivalent relationships:

dx=Qx−Px   dy=Qy−Py

or

Qx=Px+dx   Qy=Py+dy

This system is the one used by the Sinclair ZX-Spectrum and is available in BBC and Microsoft Basic.

The mathematical idea of **absolute** polar coordinates uses a fixed direction (**axis**) with a fixed point O on it as **origin.** Points P in the plane are represented by **polar coordinates** (Pr, Pa) consisting of the distance Pr of the P from O and the angle Pa between the line PO and the axis. This form can easily be converted into cartesian form and vice versa using:

x=Pr * COS(Pa)   y=Pr * SIN(Pa)

or

Pr=SQR(x ↑ 2+y ↑ 2)   Pa=ATN(y/x)

This form is not usually used directly in microcomputer implementations but an adaptation of it to define line-segments PQ in terms of the **relative** polar coordinates of Q from P is the basis of the **turtle-geometry** system used in the language LOGO and, for example, in the UCSD version of Pascal for the Apple. In this system a "turtle" at point P has an absolute Cartesian position (Px, Py) and a **heading** Pa. A line-segment PQ is defined by the distance d to be travelled from P in the direction of the current heading Pa. Thus conversion between the systems can be performed by observing that:

Qx=Px+d * COS(Pa)   Qy=Py+d * SIN(Pa)
dx=d * COS(Pa)   dy=d * SIN(Pa)

Thus conversion between the three common screen geometries is fairly easily performed.

1.3 DRAWING COMMANDS

We shall show how to draw the line-segment PQ in a number of different microcomputer dialects. First we usually need a command to clear the screen (and probably select a graphics mode).

1.3.1 Setting graphics

The following table shows some of the commands to select (and clear) a typical graphics screen in a number of different dialects:

BBC Basic:	MODE 1	selects 320×256 pixels, 4 colours
ZX-Spectrum:	CLS	selects 256×176 pixels, 8 colours
Apple II:	HGR	selects 280×160 pixels, 4 colours
RML 380Z:	CALL "RESOLU-TION",0,2	selects 320×192 pixels, 4 colours
RM Nimbus:	SET MODE 40	selects 320×250 pixels, 16 colours
Apple Mac:	CLS	selects 491×254 pixels, 2 colours
IBM PC:	SCREEN 1	selects 320×200 pixels, 4 colours

1.3.2 Moving to a point

In order to specify the position of the point P we need to know the screen coordinate system. With the BBC and RML versions of Basic the screen origin is initially at the bottom left-hand corner, but may easily be moved; with the ZX-Spectrum it is always at the bottom left corner and with the Apple II, Mac and IBM it is at the top left-hand corner. The BBC micro uses the same screen dimensions (1280×1024) irrespective of the number of pixels in the MODE selected whereas the Apple II, Mac, IBM, Spectrum and RML 380/480Z systems use the number of pixels as the screen dimensions.

To show the equivalences between the system suppose we wish to move to the first point P in the top right-hand corner of the screen about three-quarters the way along and three-quarters the way up.

BBC Basic: MOVE 960,768 moves without plotting
or: PLOT 69,960,768 plots the point
[PLOT 69, is one of some 100 different commands]
ZX-Spectrum: PLOT 192,132 plots
Apple II HPLOT 210,40 plots
[remember the origin was at the **top** left-hand corner]
RML 380Z: CALL "PLOT",240,144,3 plots
[the last ",3" is needed to plot the point in a colour]
RM Nimbus: POINTS 240, 188 plots
Apple Mac: PSET (360,60) plots
IBM PC: PSET (240,50) plots

1.3.3 Drawing a line

Again, to illustrate the equivalences between the systems, we shall now draw a horizontal line from P to a point Q one-quarter of the way in from the left-hand edge.

BBC Basic: DRAW 320,768 draws an **absolute** line
or: PLOT 1,−640,0 draws a **relative** line
[PLOT 1, is one of some 100 different commands]
ZX-Spectrum: DRAW −128,0 draws a **relative** line

Apple II:	HPLOT TO 70,40	draws an **absolute** line
[remember the origin was at the **top** left-hand corner]		
RML 380Z:	CALL "LINE",80,144	draws an **absolute** line
RM Nimbus:	LINE 240, 188; 80, 188	draws an **absolute** line
[both endpoints **have** to be given]		
Apple Mac:	Line −(120,60)	draws an **absolute** line
	LINE STEP (−240,0)	draws a **relative** line
IBM PC:	LINE −(80,50)	draws an **absolute** line

1.3.4 Drawing a closed shape

As a final comparison between the various systems we shall give sets of commands that should draw something like a right-angled triangle with sides parallel to the edges of the screen and with its bottom left corner in the middle of the screen:

```
REM - BBC Basic, absolute
MODE 1
MOVE 640,512 : DRAW 840,512 : DRAW 840,672 : DRAW 640,512


REM - BBC Basic, relative
MODE 1
MOVE 640,512 : PLOT 1,200,0 : PLOT 1,0,160 : PLOT 1,-200,-160


REM - ZX-Spectrum, relative
CLS
PLOT 128,88 : DRAW 50,0 : DRAW 0,40 : DRAW -50,-40


REM - Apple II, absolute
HGR
HPLOT 140,80 TO 190,80 TO 190,40 TO 140,80


REM - RML 380/480Z Basic, absolute
CALL "RESOLUTION",0,2
CALL "PLOT",160,96,3 : CALL "LINE",210,96
CALL "LINE",210,136 : CALL "LINE",160,96

REM - RM NIMBUS Basic, absolute
SET MODE 40
LINE 160, 125; 210, 125; 210, 165; 160, 125


REM - Apple Mac
CLS
PSET (240,120) : LINE -(340,120)
LINE -(340,80) : LINE -(240,120)


REM - IBM PC
SCREEN 1
PSET (160,100) : LINE -(210,100)
LINE -(210,60) : LINE -(160,100)
```

You should see that the hypotenuse is not really a straight line but consists of those pixels nearest to where a straight line would pass.

Now try to draw a picture of Pythagoras' theorem — i.e. put a square on each of the three sides of the triangle.

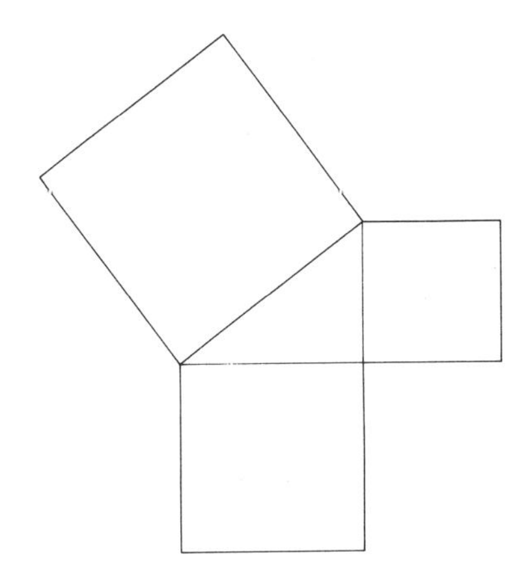

Fig. 1.1

Try drawing some designs of your own. Can you draw a hexagon?

1.3.5 Changing colour

When you select a graphics mode you are usually given two "default" choices of colour, one for the background and one for the foreground. Each colour will be given a code number. For example in MODE 1 the BBC micro selects while (code 3) for the foreground and black (code 0) for the background; the ZX-Spectrum selects black (code 0) for the foreground and white (code 7) for the background; and the RML 380Z selects black (code 0) for both! Most micros use the concept of the "current pen colour" (or "ink", or "brush", etc.) and provide means of changing it:

BBC Basic:	GCOL 0,2	2=yellow from codes 0–3
ZX-Spectrum:	INK 6	6=yellow from codes 0–7
Apple II:	HCOLOR=1	1=red from codes 0–3
RML 380Z:	CALL "PLOT",80,144,2	2=green from codes 0–3
RM Nimbus:	SET BRUSH 6	6=brown from codes 0–15
Apple Mac:	monochrome only	
IBM PC:	PSET (50,70),2	2=red from codes 0–3

Thus in BBC MODE 1 to 'rub out' an incorrect line you can change the ink colour to black with GCOL 0,0 then redraw the line and put the ink colour back to white with GCOL 0,3.

1.4 WRITING PROGRAMS

From now on the text will refer to Basic programs written in a simple form of BBC Basic. From the above discussion it should be clear that conversions between the graphics systems of different micros are not as difficult as might be expected. The last section in this chapter shows how such programs can be easily converted into equivalent programs for a range of popular micros, to emphasise these points about the ease of conversion. As the instructions you have typed eventually come further down the screen the greater are the chances that some of the text will obliterate part of your drawing. One cure for this is to store the instructions within the computer as a program — in Basic this usually just means that each line you type will be given a number to instruct the computer to store it away rather than to perform the instruction immediately. First, though, we need to ensure that the computer has no instructions already stored. The command to clear away any unwanted previously stored commands is NEW. The following program will draw some red axes across the screen:

```
100 MODE 1
110 GCOL 0,1
120 MOVE 0,512
130 DRAW 1279,512
140 MOVE 640,1023
150 DRAW 640,0
```

The numbers given to each line are deliberately chosen with gaps between to allow extra lines to be inserted, if needed. To make the program work just type:

```
RUN
```

You should find that your typing has disappeared and a picture has been drawn. If this has not happended there may be a mistake in your typing. To retrieve the listing of the program that you typed just enter:

```
LIST
```

If you want to alter a line just retype it with the original line-number. For example to change the drawing to yellow type:

```
110 GCOL 0,2
```

and LIST again to check that the change has been made. Run again to show that it works.

If you want to insert an additional line then just give it a number between the two existing lines that you want it to go. Thus to have one line in yellow and the other in red just type:

```
125 GCOL 0,1
```

and LIST and RUN again to check that things work as expected. To delete a line just type the line-number of the line to go and press **RETURN.** Guess what will happen if we remove line 140. Then type:

```
140
```

and LIST and RUN to see if you were right.

If your line numbers start to get in a bit of a mess you can tidy them up with the RENUMBER command. Try typing:

```
RENUMBER 100,10
```

and LIST to confirm that the numbers start at 100 and go up in 10s.

1.5 SOME USEFUL PROGRAMMING TECHNIQUES

1.5.1 Definite loops

This short program draws some vertical lines:

```
100 MODE 1
110 FOR X = 0 TO 1280 STEP 200
120  MOVE X,0
130  DRAW X,1024
140 NEXT X
150 END
```

Here the variable X is first of all given the value 0, then 200, then 400, and so on. The last value of X for which a line will be drawn is 1200. On adding a further 200 the computer checks that the new value of 1400 is greater than the upper limit of the loop and so goes to the line immediately after the "NEXT X", which ends the program. The spaces at the start of lines 120 and 130 are optional but have been put in to emphasise the loop structure.

In BBC Basic the programmer is not restricted to having to use simple letters such as "X" for the names of variables, but other micros are more restrictive. For example, the ZX-Spectrum only allows single-letter variable names in FOR and NEXT statements.

We can add another loop in a different way to draw some horizontal lines:

```
150 FOR count = 1 TO 10
160  ycoord = 100 * count
170  MOVE 0, ycoord : DRAW 1280, ycoord
180 NEXT count
190 END
```

Note how two instructions have been placed on the same line, separated by a colon, in line 170. If the STEP part of the FOR instruction is omitted then a step of 1 is assumed. The END statement is optional.

1.5.2 Conditional statements

If we wanted the horizontal lines to be in alternate colours then we could test to see whether the variable "count" is even or odd. The usual computer test for this is to see if there is any remainder on dividing "count" by 2. Two alternative forms of this text are:

```
155 IF count = 2* (INT(count/2)) THEN GCOL 0,1 ELSE GCOL 0,2
```

or

```
155 IF (count MOD 2) = 0 THEN GCOL 0,1 ELSE GCOL 0,2
```

The Basic function INT takes the whole number ("integral") part of its argument. The Basic operator MOD finds the remainder on division. Note that the ELSE part of the IF statement is optional. Just for completeness there is an alternative direct way of doing this job:

```
155 GCOL 0,1 + (count MOD 2)
```

1.5.3 Tested loops

In some cases we do not know in advance how many times a set of operations are to be performed but want them to be performed until some condition is satisfied. Another way to perform the job of drawing the horizontal lines in lines 110 to 140 might be:

```
110 X = 0
115 REM top of the loop
120  MOVE X,0
130  DRAW X,1024
135  X = X + 200
140 IF X<=1280 THEN GOTO 115
```

This uses the much abused GOTO statement to transfer control to the top of the loop. Notice how the "greater-than-or-equal-to" relation is written. The REM statement in line 115 just allows us to make some remarks that the computer will ignore.

With better versions of Basic, such as BBC, there is a neater way of writing this structure:

```
110 X = 0
115 REPEAT
120  MOVE X,0
130  DRAW X,1024
135  X = X + 200
140 UNTIL X>1280
```

REPEAT-UNTIL structures will be used for conditional loops in the rest of this text and the previous example shows how they can be rewritten for other micros such as the ZX-Spectrum. Microsoft Basic for the Apple Mac and IBM Basic have a similar structure called "WHILE-WEND":

```
100 CLS
110 X = 0
120 WHILE X<320
130   LINE (X,0)-(X,200)
140   X = X + 40
150 WEND
```

1.5.4 Subroutines and procedures

Often the structure of a program is made clearer if chunks of program are written as subprograms and listed in a sort of "appendix". The crudest of these structures is the "subroutine":

```
100 MODE 1
110 FOR X = 0 TO 1000 STEP 200
120   GOSUB 500 : REM draw a box
130 NEXT X
140 END
150 :
500 REM subroutine to define a box
510   MOVE X,X : DRAW X+150,X
520   DRAW X+150,X+150 : DRAW X,X+150
530   DRAW X,X
540 RETURN
```

Note how the END in line 140 is essential now to emphasise the demarcation between the main program and the appendix. The indenting of the "body" of the subroutine is optional and is done to make the structure clearer, as is the blank line in line 150.

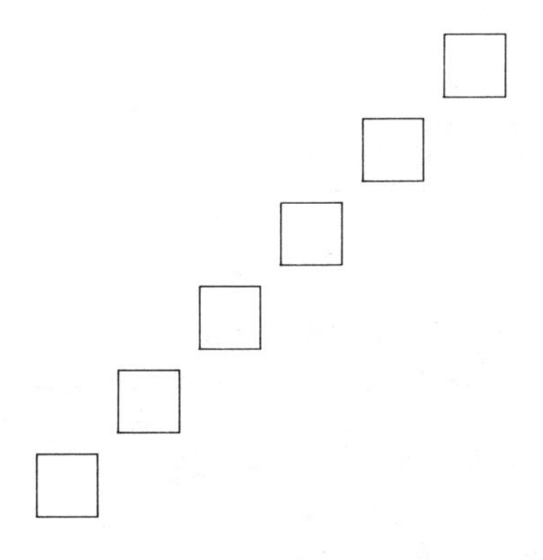

Fig. 1.2

Again, better Basics include the "procedure" as an alternative structure:

```
100 MODE 1
110 FOR X = 0 TO 1000 STEP 200
120   PROCbox
130 NEXT X
140 END
150 :
500 DEF PROCbox
510   MOVE X,X : DRAW X+150,X
520   DRAW X+150,X+150 : DRAW X,X+150
530   DRAW X,X
540 ENDPROC
```

Procedures will be widely used throughout the text but, again, the previous example shows how they can be easily rewritten in the form of subroutines. Procedures usually permit the passing of **parameters**:

```
100 MODE 1
110 FOR X = 0 TO 1000 STEP 200
120   PROCbox(X,X,150,150)
130 NEXT X
140 END
150 :
500 DEF PROCbox(x,y,w,h)
510   MOVE x,y : DRAW x+w,y
520   DRAW x+w,y+h : DRAW x,y+h
530   DRAW x,y
540 ENDPROC
```

Note how this enables us to write a general procedure for a rectangle of width w and height h with bottom left-hand corner at (x,y) and then to use it to draw a square. The syntax of procedures varies between versions, the following is the equivalent for the Apple Mac:

```
CLS
FOR X=0 TO 400 STEP 50
  BOX X,250-X/2,20.0,20.0
NEXT X
END

SUB BOX(X,Y,W,H) STATIC
  LINE (X,Y)-(X+W,Y)
  LINE -(X+W,Y-H)
  LINE -(X,Y-H)
  LINE -(X,Y)
END SUB
```

However, both Apple Mac and IBM PC have a version of the **LINE** command to draw a box (in outline, or filled):

LINE (X, Y)−(X+W,Y+H),,B

1.5.5 Nested loops

Now that we have defined PROCbox we shall not repeat its definition in each of the following programs, but you will need to include its definition after line 180 if the programs are to work! The first example uses two consecutive loops:

```
100 MODE 1
110 FOR X=0 TO 900 STEP 100
120  PROCbox(X,0,100,100)
130 NEXT X
140 FOR Y=0 TO 900 STEP 100
150  PROCbox(0,Y,100,100)
160 NEXT Y
170 END
180 :
```

How many boxes would you expect each loop to draw? How many boxes have been drawn? Can you account for any difference? What would happen if the X loop comes AFTER the Y loop?

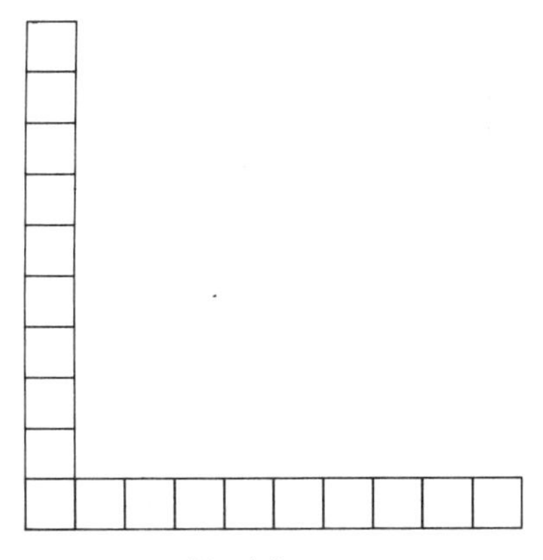

Fig. 1.3

Instead of putting the loop sequentially we could try putting one inside the other:

```
100 MODE 1
110 FOR Y=0 TO 900 STEP 100
120  FOR X=0 TO 900 STEP 100
130   PROCbox(X,Y,100,100)
140  NEXT X
150 NEXT Y
160 END
170 :
```

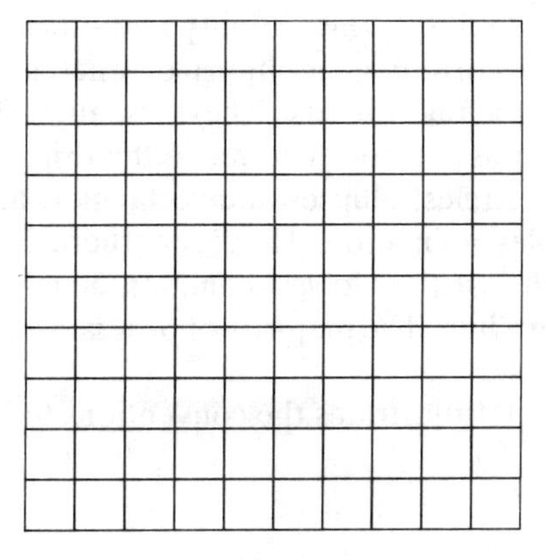

Fig. 1.4

Putting loops inside loops is known as "nesting"!

Try putting the X loop inside the Y loop. Is the picture different? Is it drawn differently? How many squares have been drawn? What is the effect of changing line 120 to:

```
120 FOR X = 0 TO Y STEP 100
```

As a final example of a variation on this theme what pattern will this program produce?

```
100 MODE 1
110 FOR X=0 TO 900 STEP 80
120  Y=X
130  PROCbox(X,Y,120,40)
140  PROCbox(X+80,Y+40,40,40)
150 NEXT X
160 END
170 :
```

1.6 THE STANDARD PROGRAM

Different microcomputers vary considerably in their versions of programming languages such as Basic. However, the way they choose to perform graphics varies more widely still. In the professional field there have been attempts to agree a common graphics standard — the Graphics Kernel System (GKS) is one such. However, we are a long way from that in respect of micros.

This is a book about geometry and computers in general; it is not about specific micros and "how to get the best from your ORANGE NUCLEON". In order to concentrate on the mathematics and to avoid the clutter of machine-dependent graphics features a compromise has to be made between readability, convertibility and efficiency. There is no doubt that some of the current generations of micros are much better suited for graphical work than others. These usually offer lots of useful additional features such as the ability to move the origin, to draw (and fill) standard shapes, like circles, ellipses and rectangles, and to apply colour in a way which enables animation. However, there is a vast amount that we can achieve with just the "lowest common denominator" of colouring a point and drawing a line. The majority of programs in this book only require those two functions.

Program 1.1 illustrates the conventions to be adopted in the rest of this book:

```
 10 REM Prog.1.1 - to draw an ellipse
 49 :
 50 MODE 1
 60 PROCsetup
 70 OX = 0.5*SW : OY = 0.5*SH
 80 SX = 100 : SY = 100
 99 :
100 PROCdot(0,0,FC)
110 A = 3 : B = 2
120 X1 = A : Y1 = 0
130 FOR T = 0 TO 6.3 STEP PI/10
140  X = A*COS(T)
150  Y = B*SIN(T)
160  PROCjoin(X1,Y1,X,Y,FC)
170  X1 = X : Y1 = Y
180 NEXT T
490 END
499 :
500 DEF PROCsetup
510  SW = 1280 : SH = 1024
520  NC = 3 : FC = 3
590 ENDPROC
599 :
600 DEF PROCdot(X,Y,PC)
610  GCOL 0,PC
620  PLOT 69, OX + X*SX, OY + Y*SY
690 ENDPROC
699 :
700 DEF PROCjoin(X1,Y1,X,Y,PC)
710  GCOL 0,PC
720  MOVE OX + X1*SX, OY + Y1*SY
730  DRAW OX + X*SX, OY + Y*SY
790 ENDPROC
799 :
```

The lines followed only by colons are just there to aid readability and may quite well be omitted. The bodies of loops and procedures are also indented with spaces, but these are optional.

The first few lines (from 10 to 49) describe the program and will usually contain a chapter and program number in line 10:

```
10 REM Prog.1.1 - to draw an ellipse
49 :
```

The next few lines (from 50 to 99) contain information which is needed to tailor the picure to a given type of micro.

```
50 MODE 1
60 PROCsetup
70 OX = 0.5*SW : OY = 0.5*SH
80 SX = 100 : SY = 100
99 :
```

Line 50 sets and clears the graphics screen.

Line 60 calls a routine to define the size of the screen, the number of colours and other machine-dependent features. For many micros this will be replaced by:

```
60 GOSUB 500
```

Line 70 sets the coordinates of the current screen origin (OX,OY), which will often be in the centre of the screen!

Line 80 sets the stretch factors for each axis SX, SY to be used to convert the coordinates of points to suitable values for the particular screen size. These will be given as ones which work for the BBC micro screen size of 1280×1024 and it is usually a matter of simple arithmetic to adapt these for any other size. Thus a common screen size of 320×200 will need factors about one-fifth as big.

The next batch of lines (from 100 to 499) contains the program. As far as possible the only graphics commands that will be included here are calls to the basic routines to **plot** a point and **draw** a line. The coordinate system used within this part of the program is quite arbitrary. Coordinates will be stretched and translated to fit on the screen according to the factors specified in line 70 and line 80.

```
100 PROCdot(0,0,FC)
110 A = 3 : B = 2
120 X1 = A : Y1 = 0
130 FOR T = 0 TO 6.3 STEP PI/10
140  X = A*COS(T)
150  Y = B*SIN(T)
160  PROCjoin(X1,Y1,X,Y,FC)
170  X1 = X : Y1 = Y
180 NEXT T
490 END
499 :
```

Line 100 will plot the current origin in the foreground colour FC. This will often need to be replaced by:

```
100 GOSUB 600
```

Line 120 sets the coordinates of the first point (X1,Y1) to be used later in line 160.

Line 130 refers to the constant PI. This is provided in some versions of Basic. Otherwise it will have to be defined in PROCsetup, e.g.:

```
530 PI = 3.14159
```

Lines 140 and 150 compute the coordinates of the next point (X,Y) to be used in line 460.

Line 160 draws the line from (X1,Y1) to (X,Y) in the foreground colour FC. This will often need to be replaced by:

```
160 GOSUB 700
```

Line 170 puts the coordinates of the last point used into (X1,Y1) ready for the next graphics command.

Apart from the changes to lines 100 and 160 this main body of the program should be transferable between different types of computers.

The next batch of lines (from 500 to 999) is used to define the main graphics tools to be used throughout the book. These will usually be omitted in what follows so it is important that you are able to add the versions needed for your type of micro.

The first routine (lines 500 to 599) defines the characteristics of the screen:

```
500 DEF PROCsetup
510   SW = 1280 : SH = 1024
520   NC = 3 : FC = 3
590 ENDPROC
599 :
```

Line 500 will often need to be replaced by:

```
500 REM PROCsetup
```

and line 590 by:

```
590 RETURN
```

Line 510 specifies the screen's width and height, SW and SH.

Line 520 specifies the number of colours NC (other than the background) that are available, and the code FC of the usual foreground colour.

Other lines may be needed to define constants such as PI.

The routine between lines 600 and 699 defines how a point on the screen is plotted. As previously, lines 600 and 690 might need amending.

```
600 DEF PROCdot(X,Y,PC)
610   GCOL 0,PC
620   PLOT 69, OX + X*SX, OY + Y*SY
690 ENDPROC
699 :
```

Line 610 selects the appropriate "pen-colour" PC.

Line 620 calcuates the screen coordinates corresponding to (X,Y) and plots the point.

The other main routine (lines 700 to 799) defines how a line is drawn from the most recent "previous" point (X1,Y1) to the "current" point (X,Y):

```
700 DEF PROCjoin(X1,Y1,X,Y,PC)
710   GCOL 0,PC
720   MOVE OX + X1*SX, OY + Y1*SY
730   DRAW OX + X*SX, OY + Y*SY
790 ENDPROC
799 :
```

Later we may have reason to add the definitions of some other useful routines, such as drawing circles and ellipses. In fact, the sample program above provides us with the means of doing just that. If we can define a routine to draw an ellipse specified by its centre, major and minor axes and colour!

PROCellipse(XC,YC,RA,RB,PC)

then we can use it to draw a circle of radius RC by making both RA and RB equal to RC:

```
800 DEF PROCcircle(XC,YC,RC,PC)
810  PROCellipse(XC,YC,RC,RC,PC)
890 ENDPROC
899 :
```

To define the routine we just need to convert lines 110 to 180 into a procedure (or subroutine):

```
900 DEF PROCellipse(XC,YC,RA,RB,PC)
910  X1 = XC + RA : Y1 = YC
920  FOR P = 0 TO 6.3 STEP PI/10
930   X = XC + RA*COS(P)
940   Y = YC + RB*SIN(P)
950   PROCjoin(X1,Y1,X,Y,PC)
960   X1 = X : Y1 = Y
970  NEXT P
990 ENDPROC
999 :
```

Lines 920 to 970 use the variable P as the "parameter" for the loop.

To test these new routines we can show how an ellipse lies between circles:

```
 10 REM Prog.1.2 - circles and ellipse
 49 :
 50 MODE 1
 60 PROCsetup
 70 OX = 0.5*SW : OY = 0.5*SH
 80 SX = 100 : SY = 100
 99 :
100 PROCdot(0,0,FC)
110 A = 3 : B = 2
120 PROCcircle(0,0,A,1)
130 PROCcircle(0,0,B,2)
140 PROCellipse(0,0,A,B,FC)
490 END
```

Any other general purpose routines developed in the book will have four-figure line numbers (from 1000 to 9999). Thus we could define a routine to draw a box of width BW and height BH with (X,Y) as its bottom left-hand corner by:

```
1000 DEF PROCbox(X,Y,BW,BH,PC)
1010  PROCjoin(X,     Y,      X+BW,Y,     PC)
1020  PROCjoin(X+BW,Y,      X+BW,Y+BH,  PC)
1030  PROCjoin(X+BW,Y+BH,  X,     Y+BH,  PC)
1040  PROCjoin(X,     Y+BH,  X,     Y,     PC)
1090 ENDPROC
1099 :
```

Where there is cause to have some machine-specific routines that are not widely transferable then these will have five-figure line numbers. For example the following routine, to fill the triangle (X2,Y2), (X1,Y1), (X,Y) with colour, is not transferable to all micros:

```
10000 DEF PROCfill(X2,Y2,X1,Y1,X,Y,PC)
10010  GCOL 0,PC
10020  MOVE OX + X2*SX, OY + Y2*SY
10030  MOVE OX + X1*SX, OY + Y1*SY
10040  PLOT 85, OX + X*SX, OY + Y*SY
10090 ENDPROC
10099 :
```

1.7 CONVERTING PROGRAMS FOR OTHER MICROS

The list of micros for which this could now be done is quite massive, so just a few machines have been chosen to illustrate the principles involved. If your particular micro is not among these then by reading through these, together with your computer's manual, you should be able to make the necessary conversions. The chosen micros are:

A: Apple II
B+: BBC B-plus and Master 128
I: IBM PC
M: Apple MacIntosh
N: Research Machines RM Nimbus
R: Research Machines 380/480Z
S: Sinclair ZX-Spectrum

1.7.1 Converting for the Apple II

Here we have a screen size of 280×160 with colour codes 0–7. The origin is in the **top** left-hand corner and there are no procedures, so we have to use subroutines:

```
 10 REM Prog.C1.1(A) - to draw an ellipse, Apple II
 49 :
 50 HRG
 60 GOSUB 500 : REM PROCsetup
 70 OX = 0.5*SW : OY = 0.5*SH
 80 SX = 15 : SY = 15
 99 :
100 X = 0 : Y = 0 : PC = FC : GOSUB 600 : REM PROCdot(0,0,FC)
110 A = 3 : B = 2
120 X1 = A : Y1 = 0
130 FOR T = 0 TO 6.3 STEP PI/10
140  X = A*COS(T)
150  Y = B*SIN(T)
```

```
160  GOSUB 700 : REM PROCjoin(X1,Y1,X,Y,FC)
170  X1 = X : Y1 = Y
180 NEXT T
490 END
499 :
500 REM PROCsetup
510  SW = 280 : SH = 160
520  NC = 7 : FC = 3
530  PI = 3.14159
590 RETURN
599 :
600 REM PROCdot(X,Y,PC)
610  HCOLOR = PC
620  HPLOT OX+X*SX,OY-Y*SY
690 RETURN
699 :
700 REM PROCjoin(X1,Y1,X,Y,PC)
710  HCOLOR = PC
720  HPLOT OX+X1*SX,OY-Y1*SY
730  HPLOT TO OX+X*SX,OY-Y*SY
790 RETURN
799 :
```

1.7.2 Converting for the BBC B+ and Master

These machines have the same graphics commands used throughout the book. They also incorporate commands from the graphics extension ROM which provide other utilities for circles and ellipses:

```
800 DEF PROCcircle(XC,YC,RC,PC)
810  PROCdot(XC,YC,PC)
820  PLOT 145, RC*SX, 0
890 ENDPROC
899 :
900 DEF PROCellipse(XC,YC,RA,RB,PC)
910  PROCdot(XC,YC,PC)
920  PROCdot(XC+RA,YC,PC)
930  PLOT 193, -RA*SX, RB*SY
990 ENDPROC
999 :
```

1.7.3 Converting for the IBM PC

There are three versions of Basic for the IBM PC. The version called BASICA (the "A" is for advanced) includes the graphics commands, and the micro must have a graphics card installed. There are two different graphic modes. SCREEN 1 provides 320×200 points and colours 0–3. SCREEN 2 gives 640×200 points and colours 0–1. The origin is at the top left-hand corner.

```
 10 REM Prog.1.1(I) - to draw an ellipse, IBM PC
 49 :
 50 SCREEN 1
 60 GOSUB 500 : REM PROCsetup
 70 OX = 0.5*SW : OY = 0.5*SH
 80 SX = 20 : SY = 20
 99 :
100 X = 0 : Y = 0 : PC = FC : GOSUB 600 : REM PROCdot(0,0,FC)
```

```
 110 A = 3 : B = 2
 120 X1 = A : Y1 = 0
 130 FOR T = 0 TO 6.3 STEP PI/10
 140  X = A*COS(T)
 150  Y = B*SIN(T)
 160  GOSUB 700 : REM PROCjoin(X1,Y1,X,Y,FC)
 170  X1 = X : Y1 = Y
 180 NEXT T
 490 END
 499 :
 500 REM PROCsetup
 510  SW = 320 : SH = 200
 520  NC = 3 : FC = 3
 530  PI = 3.14159
 590 RETURN
 599 :
 600 REM PROCdot(X,Y,PC)
 610  PSET (OX+X*SX,OY-Y*SY),PC
 690 RETURN
 699 :
 700 REM PROCjoin(X1,Y1,X,Y,PC)
 710  PSET (OX+X1*SX,OY-Y1*SY),PC
 720  LINE -(OX+X*SX,OY-Y*SY)
 790 RETURN
 799 :
 800 REM PROCcircle(XC,YC,RC,PC)
 810  CIRCLE (OX+XC*SX,OY-YC*SY),RC*SX,PC
 890 RETURN
 899 :
 900 REM PROCellipse(XC,YC,RA,RB,PC)
 910  CIRCLE (OX+XC*SX,OY-YC*SY),RC*SX,PC,,,RB/RA
 990 RETURN
 999 :
1000 REM PROCbox(X,Y,BW,BH,PC)
1010  PSET (OX+X*SX,OY-Y*SY),PC
1020  LINE -STEP(BW*SX,-BH*SY),PC,B
1090 RETURN
1099 :
```

1.7.4 Converting for the Apple Macintosh

There is a version of Microsoft Basic for the Mac which is very similar in some respects to the IBM PC. The graphics commands are virtually identical except that the screen is 491×253 and monochrome. The system does not need line numbers, and subroutines can be labelled:

```
REM Prog.1.1(M) - to draw an ellipse, Apple Mac
CLS
GOSUB setup
OX = 0.5*SW : OY = 0.5*SH
SX = 25 : SY = 25
X = 0 : Y = 0 : PC = FC : GOSUB dot
A = 3 : B = 2
X1 = A : Y1 = 0
FOR T = 0 TO 6.3 STEP PI/10
 X = A*COS(T)
 Y = B*SIN(T)
 GOSUB join
 X1 = X : Y1 = Y
NEXT T
END
```

```
setup
 SW = 490 : SH = 250
 NC = 1 : FC = 30
 PI = 3.14159
RETURN

dot
 PSET (OX+X*SX,OY-Y*SY)
RETURN

join
 PSET (OX+X1*SX,OY-Y1*SY)
 LINE -(OX+X*SX,OY-Y*SY)
RETURN

pcircle
 CIRCLE (OX+XC*SX,OY-YC*SY),RC*SX
RETURN

ellipse
 CIRCLE (OX+XC*SX,OY-YC*SY),RC*SX,,,,RB/RA
RETURN

box
 PSET (OX+X*SX,OY-Y*SY)
 LINE -STEP(BW*SX,-BH*SY),,B
RETURN
```

1.7.5 Converting for the RM Nimbus

RM Basic provides two graphic modes. MODE 40 gives 320×250 points with colours 0–15, and MODE 80 gives 640×250 with colours 0–3.

```
 10 REM Prog.1.1(N) - to draw an ellipse, RM Nimbus
 49 :
 50 SET MODE 40
 55 GLOBAL Sw, Sh, Nc, Fc, Ox, Oy, Sx, Sy
 60 Setup
 70 Ox := 0.5 * Sw : Oy := 0.5 * Sh
 80 Sx := 20 : Sy := 20
 99 :
100 Dot 0, 0, Fc
110 A := 3 : B := 2
120 X1 := A : Y1 := 0
130 FOR T := 0 TO 6.3 STEP PI / 10
140   X := A * COS(T)
150   Y := B * SIN(T)
160   Join X1, Y1, X, Y, Fc
170   X1 := X : Y1 := Y
180 NEXT T
490 END
499 :
500 PROCEDURE Setup
505   GLOBAL Sw, Sh, Nc, Fc
510   Sw := 320 : Sh := 250
520   Nc := 15 : Fc := 15
590 ENDPROC
599 :
600 PROCEDURE Dot X, Y, Pc
605   GLOBAL Ox, Oy, Sx, Sy
610   SET BRUSH Pc
620   POINTS Ox + X * Sx, Oy + Y * Sy
```

```
  690 ENDPROC
  699 :
  700 PROCEDURE Join X, Y, X1, Y1, Pc
  705   GLOBAL Ox, Oy, Sx, Sy
  710   SET BRUSH Pc
  720   LINE Ox + X1 * Sx, Oy + Y1 * Sy; Ox + X * Sx, Oy + Y * SY
  790 ENDPROC
  799 :
  800 PROCEDURE Pcircle Xc, Yc, Rc, Pc
  805   GLOBAL Ox, Oy, Sx, Sy
  810   SET BRUSH Pc
  820   CIRCLE Rc * Sx, Ox + Xc * Sx, Oy + Yc * Sy STYLE 3
  890 ENDPROC
  899 :
10000 PROCEDURE Fill X2, Y2, X1, Y1, X, Y, Pc
10005   GLOBAL Ox, Oy, Sx, Sy
10010   SET BRUSH Pc
10020   AREA Ox + X2 * Sx, Oy + Y2 * Sy; Ox + X1 * Sx, Oy + Y1 * Sy;
        Ox + X * Sx, Oy + Y * Sy STYLE 1
10090 ENDPROC
10099 :
```

There are very useful commands to set the scales on the axes and to set the origin. Users of the Nimbus will quickly want to replace these procedures with those more suited to the sophistication of the machine.

```
 10 REM Prog.1.1(N) - to draw an ellipse, RM Nimbus
 49 :
 50 SET MODE 40, 8, 6
 60 SET ORIGIN 4, 3
 99 :
100 POINTS 0, 0
110 A := 3 : B := 2
120 X1 := A : Y1 := 0
130 FOR T := 0 TO 6.3 STEP PI / 10
140   X := A * COS(T)
150   Y := B * SIN(T)
160   LINE X1, Y1; X, Y
170   X1 := X : Y1 := Y
180 NEXT T
490 END
```

1.7.6 Converting for the Research Machines 380/480Z

There are two graphics modes, one of which has 320×192 screen and colours coded 0–3.

```
 10 REM Prog.1.1(R) - to draw an ellipse, RML 480Z
 49 :
 50 CALL "RESOLUTION",0,2
 60 GOSUB 500 : REM PROCsetup
 70 OX = 0.5*SW : OY = 0.5*SH
 80 SX = 25 : SY = 25
 99 :
100 X = 0 : Y = 0 : PC = FC : GOSUB 600 : REM PROCdot(0,0,FC)
110 A = 3 : B = 2
120 X1 = A : Y1 = 0
130 FOR T = 0 TO 6.3 STEP PI/10
140  X = A*COS(T)
150  Y = B*SIN(T)
160  GOSUB 700 : REM PROCjoin(X1,Y1,X,Y,FC)
```

```
170  X1 = X : Y1 = Y
180 NEXT T
490 END
499 :
500 REM PROCsetup
510  SW = 320 : SH = 192
520  NC = 3 : FC = 3
530  PI = 3.14159
590 RETURN
599 :
600 REM PROCdot(X,Y,PC)
610  CALL "PLOT",OX+X*SX,OY+Y*SY,PC
690 RETURN
699 :
700 REM PROCjoin(X1,Y1,X,Y,PC)
710  CALL "PLOT",OX+X1*SX,OY+Y1*SY,PC
720  CALL "LINE",OX+X*SX,OY+Y*SY
790 RETURN
799 :
```

1.7.7 Converting for the Sinclair ZX-Spectrum

The graphics screen has 256×176 points and colours coded 0–7.

```
 10 REM Prog.1.1(S) - to draw an ellipse, ZX-Spectrum
 49 :
 50 CLS
 60 GO SUB 500 : REM PROCsetup
 70 LET OX = 0.5*SW : LET OY = 0.5*SH
 80 LET SX = 20 : LET SY = 20
 99 :
100 LET X = 0 : LET Y = 0 : LET PC = FC
105 GO SUB 600 : REM PROCdot(0,0,FC)
110 LET A = 3 : LET B = 2
120 LET X1 = A : LET Y1 = 0
130 FOR T = 0 TO 6.3 STEP PI/10
140 : LET X = A*COS (T)
150 : LET Y = B*SIN (T)
160 : GO SUB 700 : REM PROCjoin(X1,Y1,X,Y,FC)
170 : LET X1 = X : LET Y1 = Y
180 NEXT T
490 END
499 :
500 REM PROCsetup
510 : LET SW = 256 : LET SH = 176
520 : LET NC = 7 : LET FC = 0
590 RETURN
599 :
600 REM PROCdot(X,Y,PC)
610 : INK PC
620 : PLOT OX+X*SX,OY+Y*SY
690 RETURN
699 :
700 REM PROCjoin(X1,Y1,X,Y,PC)
710 : INK PC
720 : PLOT OX+X1*SX,OY+Y1*SY
730 : DRAW (X-X1)*SX,(Y-Y1)*SY
790 RETURN
799 :
800 REM PROCcircle(XC,YC,RC,PC)
```

```
810 : CIRCLE OX+XC*SX,OY+YC*SY,RC*SX
890 RETURN
899 :
```

1.7.8 Speeding up on the BBC

The programs in the book will run on the BBC micro, but have been written to make conversion for other micros as easy as possible. The present program would run much more quickly if we did away with the calls to drawing procedures, used the graphics commands directly, and took advantage of the "change of origin" command: VDU 29,OX;OY;

```
 10 REM Prog.1.1 - to draw an ellipse
 49 :
 50 MODE 1
 60 VDU 29,640;512;
 80 SX = 100 : SY = 100
 99 :
100 PLOT 69,0,0
110 A = 3 : B = 2
120 X1 = A : Y1 = 0
125 MOVE X1*SX,Y1*SY
130 FOR T = 0 TO 6.3 STEP PI/10
140  X = A*COS(T)
150  Y = B*SIN(T)
160  DRAW X*SX,Y*SY
180 NEXT T
490 END
```

2

The circle line

2.1 A LEANING LADDER

Suppose we want to draw a white 'ladder' with its foot on the x-axis and its top against the y-axis. if the length of the ladder is L and the foot of the ladder is on the ground at (G,0), referred to the centre of the screen as origin, then the top of the ladder is against the wall at (0,W) where G↑2+W↑2=L↑2. To write W in terms of G involves taking a square root — one way of putting this in Basic is: W=SQR(L*L−G*G) where the '*' symbol means multiplication. Thus a pair of red axes and a single white ladder can be drawn with the commands:

```
 10 REM Prog.2.1 - A single ladder
 49 :
 50 MODE 1 : PROCsetup
 60 OX = 0.5*SW : OY = 0.5*SH
 70 SX = 100 : SY = 100
 99 :
100 PROCjoin(5.0, 0.0, 1)
110 PROCjoin(0.0, 0.5, 1)
120 L = 4
130 G = 1
140 W = SQR(L*L-G*G)
150 PROCjoin(G.0, 0.W, 3)
490 END
499 :
500 DEF PROCsetup
510   SW = 1280 : SH = 1024
520   NC = 3 : FC = 3
590 ENDPROC
599 :
700 DEF PROCjoin(X1,Y1,X,Y,PC)
710   GCOL 0,PC
720   MOVE OX + X1*SX, OY + Y1*SY
730   DRAW OX + X*SX, OY + Y*SY
790 ENDPROC
799 :
```

Remember to put the versions of PROCsetup and PROCjoin that are appropriate to your own make of micro.

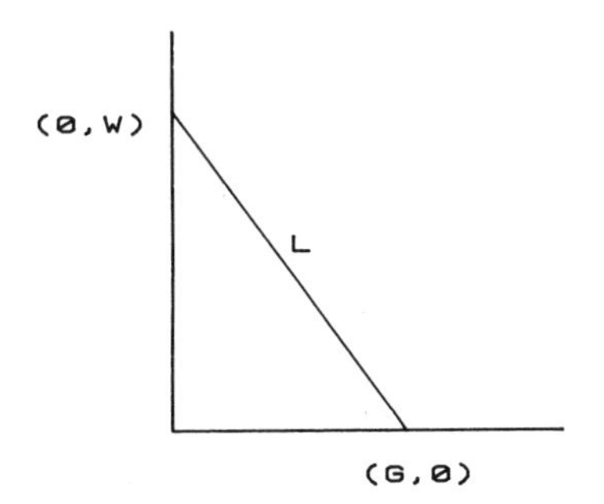

Fig. 2.1 — A single ladder.

2.2 LOTS OF LEANING LADDERS MAKE AN ENVELOPE

It is a very simple matter, now, to draw lots of ladders. We just need to be able to repeat the whole process for a range of suitable choices of G. In the last chapter we met a simple way of doing this called a FOR-NEXT loop. Suppose we want to use values of G starting at 0 (i.e. a vertical ladder), and finishing at G=L (i.e. a horizontal ladder). If we want, say, about ten different positions of the ladder to be displayed then we need G to increase in steps of L/10. The required additions are:

```
 10 REM Prog.2.2 - Lots of ladder
 49 :
 50 MODE 1 : PROCsetup
 60 OX = 0.5*SW : OY = 0.5*SH
 70 SX = 100 : SY = 100
 99 :
100 PROCjoin(5,0, 0,0, 1)
110 PROCjoin(0,0, 0,5, 1)
120 L = 4
130 FOR G = 0 TO L STEP L/10
140  W = SQR(L*L-G*G)
150  PROCjoin(G,0, 0,W, 3)
160 NEXT G
490 END
```

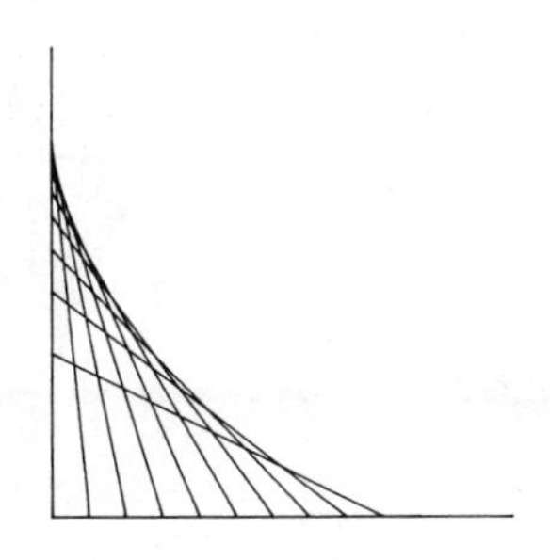

Fig. 2.2 — Ladders in the first quadrant.

This idea of having a small amount of program steps obeyed many times in a **loop** is one of the most frequently used and powerful components of computer programs. To have twice as many ladders we need only alter the value of the STEP used in line 150 from L/10 to L/20. See what happens if you use a step length of L/100.

In order to make the ladder 'climb' from the horizontal up to the vertical we just need to count the values of X in the other direction:

```
130 FOR G=L TO 0 STEP -L/10
```

Rather than have a falling ladder we can easily adapt the program to produce a 'falling' rhombus. After positioning the 'pen' at (G,0) and drawing to (0,W) we need only continue to draw to (−G,0), then to (0,−W) and, finally, back again to (G,0):

```
160  PROCjoin(0,W, -G,0, 3)
170  PROCjoin(-G,0, 0,-W, 3)
180  PROCjoin(0,-W, G,0, 3)
190 NEXT G
```

The set of straight lines gives an impression of a curve which is a four-pointed star. This curve is the **astroid** and the lines are all tangents to it. Thus the set of lines envelopes an astroid.

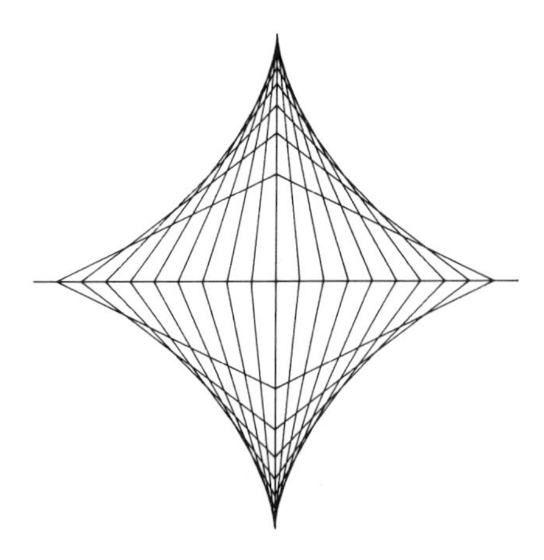

Fig. 2.3 — Envelope of astroid.

2.3 PARAMETRIC REPRESENTATION

The positions of the ladder are not very evenly spaced. By making the distance G of the foot of the ladder from the wall the **independent** variable, and the height W of the top of the ladder the **dependent** variable we have produced one set of ladders. A different approach would be to choose the angle T that the ladder makes with the floor as the variable and to calculate both G and W in terms of T using simple trigonometry:

G=L*COS(T)   W=L*SIN(T)

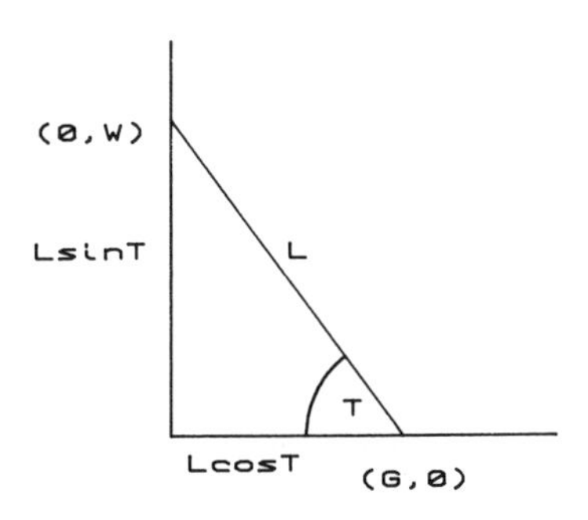

Fig. 2.4.

Thus to make the ladders start horizontal and finish vertical, drawn at intervals of 10 degrees we might try the following changes:

```
130 FOR T = 0 TO 90 STEP 10
140   G = L*COS(T) : W = L*SIN(T)
150   PROCjoin(G,0, 0,W, 3)
160 NEXT T
```

Unfortunately this does not produce the neat picture we might have imagined. The computer calculates its trigonometric functions (SIN, COS and TAN) using angles measured in **radians.** In this system there are 2*PI radians in a full circle (instead of 360 degrees). Thus the angle in a quarter circle is PI/2 radians. Fortunately the value of PI (3.14159....) is already stored for us in BBC Basic. The only change needed is:

```
130 FOR T=0 TO PI/2 STEP PI/18
```

With any luck this should now produce the desired effect.

In many applications in computer graphics it is preferable to use an angle T as the independent variable and to find the values of X and Y for screen coordinates as functions of T. This kind of representation is usually called **parametric** form and the angle T is called a **parameter.** This means, though, that we shall have to get familiar with measuring angles in radians. In fact, one radian is just slightly smaller than 60 degrees (about 57 degrees) and hence there are just a bit more than 6 of them in a circle (i.e. 2*PI). As a working aproximation we shall often use 6.3 as being quite close to 2*PI.

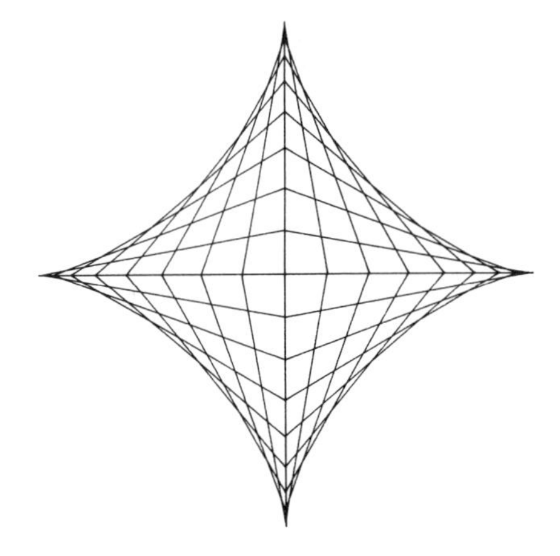

Fig. 2.5 — The envelope of an astroid with equi-spaced parameter.

2.4 TURNING AN ENVELOPE INTO A LOCUS

We can also make the computer draw 'curves' by approximating them with lots of little straight lines. For example we can explore the curve traced out by the feet of a 'painter' standing on the middle rung of the ladder as the ladder slips. The top of the ladder is at (0,W) and foot at (G,0), thus the middle of the ladder is at (G/2,W/2). In order to draw the curve we just need to draw a line from the centre of the ladder to the centre of the previous ladder:

```
 10 REM Prog.2.3 - Locus of the centre of the ladder
 49 :
 50 MODE 1 : PROCsetup
 60 OX = 0.5*SW : OY = 0.5*SH
 70 SX = 100 : SY = 100
 99 :
100 PROCjoin(5,0, 0,0, 1)
110 PROCjoin(0,0, 0,5, 1)
120 L = 4
125 X1 = L/2 : Y1 = 0
130 FOR T = 0 TO PI/2 STEP PI/18
140  G = L*COS(T) : W = L*SIN(T)
150  X = G/2 : Y = W/2
160  PROCjoin(X1,Y1, X,Y, 3)
170  X1 = X : Y1 = Y
180 NEXT T
490 END
```

You should now see a quarter-circle in the first quadrant.

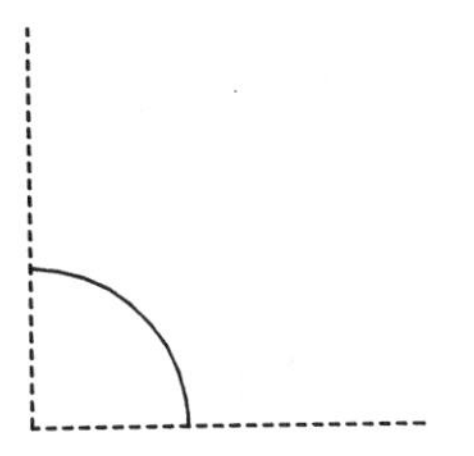

Fig. 2.6 — Quarter-circle.

By changing the values used in line 130 it should be possible to draw a complete circle. We just need to change the upper limit of 'T' from PI/2 to 2*PI. However, you should find that the curve does not quite join up. This is due to the fact that the computer cannot represent PI/18 exactly, and so when it adds it to itself 36 times it does not quite come up to the same value as 2*PI. Thus we need to use a working approximation for 2*PI such as 6.3 which is just a shade greater than 2*PI (by about a quarter of a percent) and the statement:

```
130 FOR T = 0 to 6.3 STEP PI/18
```

should do the trick. Note that we have in fact approximated the circle by a 36 point polgon. You can easily vary the step in line 180 to make polygons with more or less sides. To get a hexagon use STEP PI/3. In fact there is little sense in using more points to approximate a circle — it only slows down the drawing with no noticeable increase in accuracy. The 'industry standard circle' for this kind of resolution is a 40-point polygon!

The following program draws a 40-sided polygon of radius R with centre at (640,512):

```
 10 REM Prog.2.4 - Drawing a "circle"
 49 :
 50 MODE 1 : PROCsetup
 60 OX = 0.5*SW : OY = 0.5*SH
 70 SX = 100 : SY = 100
 99 :
100 R = 4
110 X1 = R : Y1 = 0
120 FOR T = 0 TO 6.3 STEP PI/20
130   X = R*COS(T) : Y = R*SIN(T)
140   PROCjoin(X1,Y1, X,Y, 3)
150   X1 = X : Y1 = Y
160 NEXT T
490 END
```

Fig. 2.7 — A circle.

2.5 IF POLYGONS BECOME CIRCLES — HOW DO CIRCLES BECOME ELLIPSES?

Suppose our mythical painter stood somewhere other than in the middle of the ladder, what would the curve formed by the locus of his feet look like? Instead of using the fraction 1/2 to fix the midpoint of the ladder we can introduce another parameter F to specify the fraction of the way up the ladder that the painter stands. When F=0 he is at the foot, and when F=1 he is at the top. If the foot of the ladder is at (G,0) and the top is at (0,W) then we can use similar triangles to find that, for a particular choice of F, his feet are at the point ((1−F)*G,F*W). Thus we can make slight changes to Prog. 2.3 to explore the locus of, say, a point one-third the way up the ladder:

```
 10 REM Prog.2.5 - Locus of a point on a falling ladder
 49 :
 50 MODE 1 : PROCsetup
 60 OX = 0.5*SW : OY = 0.5*SH
 70 SX = 100 : SY = 100
 99 :
100 L = 4
110 F = 1/3
120 X1 = (1-F)*L : Y1 = 0
130 FOR T = 0 TO 6.3 STEP PI/18
140   G = L*COS(T) : W = L*SIN(T)
150   X = (1-F)*G : Y = F*W
160   PROCjoin(X1,Y1, X,Y, 3)
170   X1 = X : Y1 = Y
180 NEXT T
490 END
```

We now have a very short program that can be used to draw circles, regular polygons, ellipses and "elliptical polygons" of any size in any position:

Try different values of F. What happens if F is greater than 1, or if F is negative?

If we write:

$$A = (1-F)*L \qquad \text{and} \qquad B = F*L$$

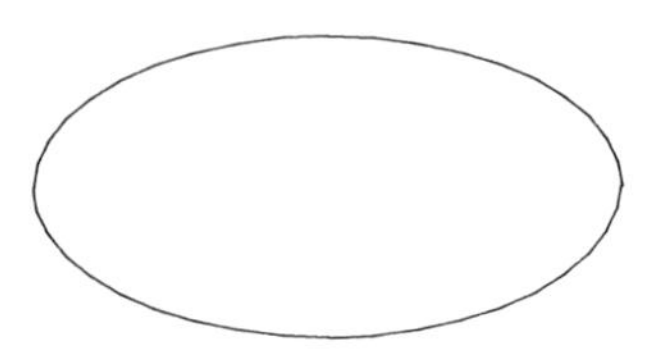

Fig. 2.8 — An ellipse.

then we have a slightly neater, but equivalent, version of this program, which we first met in section 1.6:

```
 10 REM Prog.2.6 - An ellipse
 49 :
 50 MODE 1 : PROCsetup
 60 OX = 0.5*SW : OY = 0.5*SH
 70 SX = 100 : SY = 100
 99 :
100 A = 3 : B = 2
110 X1 = A : Y1 = 0
120 FOR T = 0 TO 6.3 STEP PI/18
130  X = A*COS(T) : Y = B*SIN(T)
140  PROCjoin(X1,Y1, X,Y, 3)
150  X1 = X : Y1 = Y
160 NEXT T
490 END
```

Here are a number of different designs based on circles, polygons and ellipses. Can you adapt the program to draw each of them?

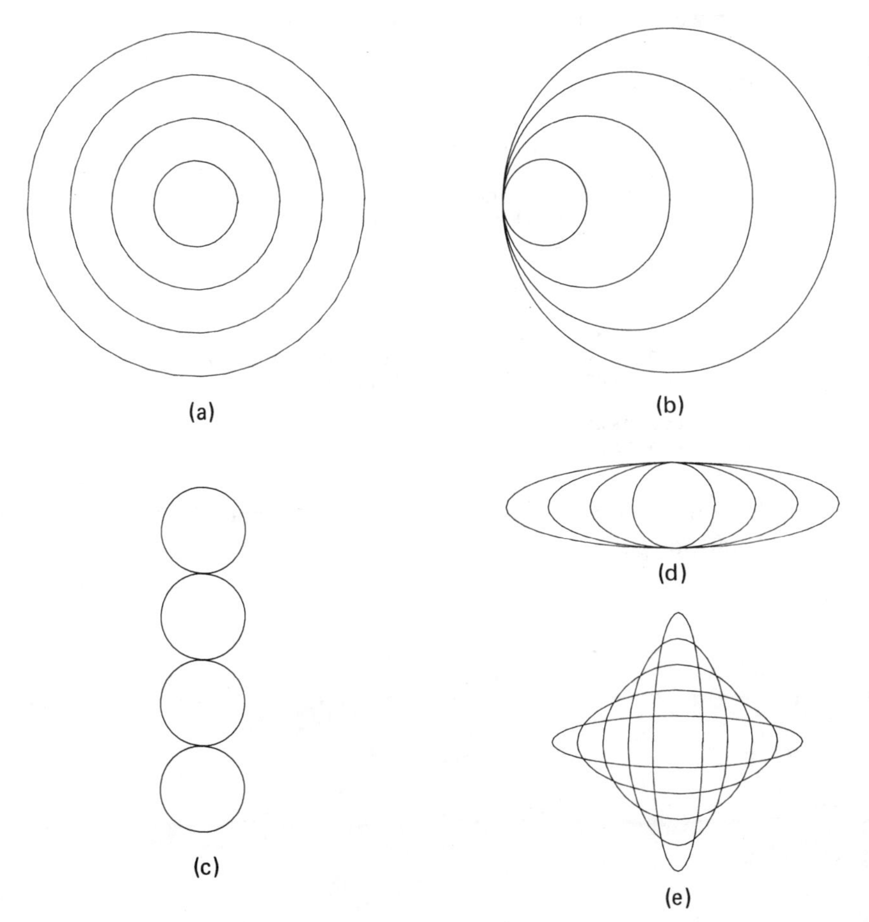

Fig. 2.9 — A set of curves.

2.6 SOME VARIATIONS ON POLYGONS

We can adapt Prog. 2.6 to make the loop count exactly the number of sides of the polygon by:

```
 10 REM Prog.2.7 - Regular polygon
 49 :
 50 MODE 1 : PROCsetup
 60 OX = 0.5*SW : OY = 0.5*SH
 70 SX = 100 : SY = 100
 99 :
100 R = 4 : N = 6 : A = 2*PI/N
110 X1 = R : Y1 = 0
120 FOR I = 1 TO N
130  X = R*COS(I*A) : Y = R*SIN(I*A)
140  PROCjoin(X1,Y1, X,Y, 3)
150  X1 = X : Y1 = Y
160 NEXT I
490 END
```

If N is not a whole number then the figure will not close up. Suppose N=2.4 then we can write N as a fraction in lowest terms as N=12/5, and if we now draw 12 sides we should get a polygon that closes up. We just need to change lines 100 and 120:

```
 10 REM Prog.2.8 - Star polygon
 49 :
 50 MODE 1 : PROCsetup
 60 OX = 0.5*SW : OY = 0.5*SH
 70 SX = 100 : SY = 100
 99 :
100 R = 4 : P = 12 : Q = 5 : N = P/Q : A = 2*PI/N
110 X1 = R : Y1 = 0
120 FOR I = 1 TO P
130  X = R*COS(I*A) : Y = R*SIN(I*A)
140  PROCjoin(X1,Y1, X,Y, 3)
150  X1 = X : Y1 = Y
160 NEXT I
490 END
```

Try using other coprime values of P and Q to investigate other "star-polygons".

Fig. 2.10 — A star polygon.

Another interesting problem concerns the "nesting" of polygons within each other. Suppose a regular N-gon is inscribed in a circle of radius R_N and a regular (N+1)-gon is inscribed in a circle of radius R_{N+1}, what is the relationship between the radii if the second circle is to touch each of the sides of the N-gon?

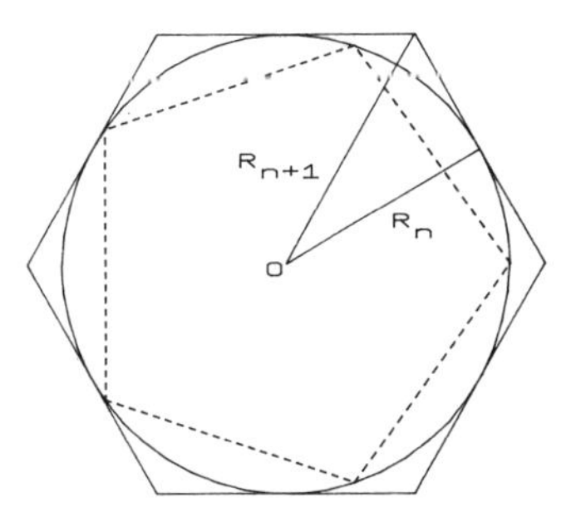

Fig. 2.11 — An n+1-gon inscribed in an n-gon.

By simple trigonometry we have that:

$$R_{N+1}=R_N * \text{COS}(\text{PI}/N)$$

Thus the next program will draw a sequence of polygons that are "nested" inside each other:

```
 10 REM Prog.2.9 - Nested polygons
 49 :
 50 MODE 1 : PROCsetup
 60 OX = 0.5*SW : OY = 0.5*SH
 70 SX = 100 : SY = 100
 99 :
100 R = 5
110 FOR N = 3 TO 20
120  A = 2*PI/N
130  X1 = R : Y1 = 0
140  FOR I = 1 TO N
150   X = R*COS(I*A) : Y = R*SIN(I*A)
160   PROCjoin(X1,Y1, X,Y, 3)
170   X1 = X : Y1 = Y
180  NEXT I
190  R = R*COS(PI/N)
200 NEXT N
490 END
```

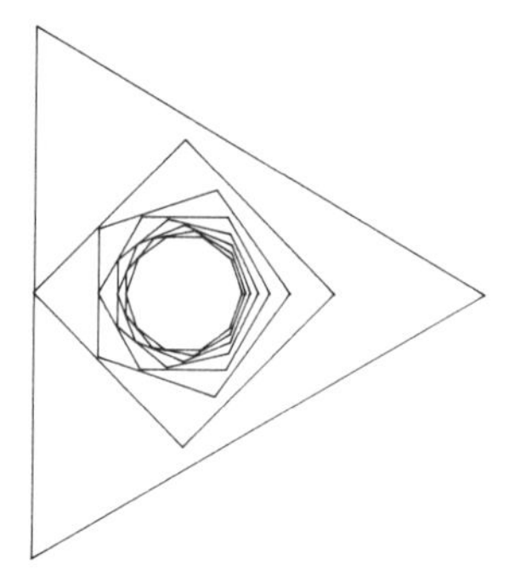

Fig. 2.12 — Plot of nested polygons.

Do you think there will always be a "hole in the middle" however large we let the value of N become? If so, how big is the hole?

2.7 CURVE STITCHING IN A CIRCLE

A well-documented investigation consists of numbering a set of equally spaced points round a circle 1,2, ..., N and chosing some multiplier M. Lines are then drawn between points 1,M; 2,2*M; 3,3*M; etc. to envelope a curve. the curve turns out to be an epicycloid with (M−1) cusps. We can easily adapt our polygon drawing program to model this:

```
 10 REM Prog.2.10 - Envelope of epicycloid
 49 :
 50 MODE 1 : PROCsetup
 60 OX = 0.5*SW : OY = 0.5*SH
 70 SX = 100 : SY = 100
 99 :
100 R = 5 : M = 4
110 N = 144
120 A = 2*PI/N
130 FOR I = 1 TO N
140  A1 = I*A : A2 = M*A1
150  X1 = R*COS(A1) : Y1 = R*SIN(A1)
160  X  = R*COS(A2) : Y  = R*SIN(A2)
170  PROCjoin(X1,Y1, X,Y, 3)
180 NEXT I
490 END
```

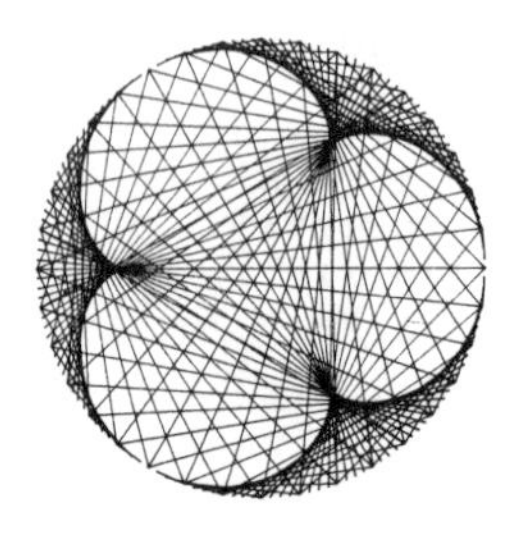

Fig. 2.13 — Trefoil eneveloped by curve stitching.

Try other values of M. Can you adapt the principle used in the "star-polygons" to cope with values of M which are not whole numbers?

2.8 ROTATIONS, TILTED ELLIPSES AND FASTER CIRCLES

How can the ellipse drawn by Prog. 2.6 be tilted so that its diameters are no longer horizontal and vertical? Suppose the ellipse is rotated through an angle PH then the coordinates of all points can be transformed using the rotation matrix:

$$\begin{pmatrix} \cos(PH) & -\sin(PH) \\ \sin(PH) & \cos(PH) \end{pmatrix}$$

Thus, if (XE,YE) is a point on an "untilted ellipse", then the corresponding point (X,Y) on the ellipse tilted through an angle PH is given by:

X=XE*COS(PH)−YE*SIN(PH)
Y=XE*SIN(PH)+YE*COS(PH)

When we come to put this in a program we note that PH stays as a constant and hence we only need to evaluate COS(PH) and SIN(PH) once. Suppose these are stored as C and S respectively then:

X=XE*C−YE*S   Y=XE*S+YE*C

and this is a relationship that we shall have frequent cause to use.

Only slight changes are needed to the original program:

```
 10 REM Prog.2.11 - A tilted ellipse
 49 :
 50 MODE 1 : PROCsetup
 60 OX = 0.5*SW : OY = 0.5*SH
 70 SX = 100 : SY = 100
 99 :
100 A = 3 : B = 2 : N = 40
110 PH = PI/6 : C = COS(PH) : S = SIN(PH)
120 X1 = A*C : Y1 = A*S
130 FOR T = 0 TO 6.3 STEP 2*PI/N
140   XE = A*COS(T) : YE = B*SIN(T)
150   X = XE*C - YE*S : Y = XE*S + YE*C
160   PROCjoin(X1,Y1, X,Y, 3)
170   X1 = X : Y1 = Y
180 NEXT T
490 END
```

You can make PH change in an outer loop to draw lots of ellipses at different

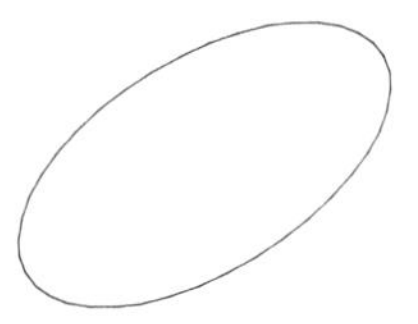

Fig. 2.14 — A tilted ellipse.

angles — what other effects can you achieve using this technique?

One application of this technique is to produce a more efficient program to draw a circle. If the circle is to be approximated by an N-sided polygon we just need to fix the starting point and then to repeatedly multiply by the matrix for a rotation of 2*PI/N radians — this means we just need to calculate one sine and one cosine which can be stored as constants S and C:

```
 10 REM Prog.2.12 - A fast circle
 49 :
 50 MODE 1 : PROCsetup
 60 OX = 0.5*SW : OY = 0.5*SH
 70 SX = 100 : SY = 100
 99 :
100 R = 4 : N = 40
110 A = 2*PI/N : C = COS(A) : S = SIN(A)
120 X1 = R : Y1 = 0
130 FOR I = 1 TO N
140  X = X1*C - Y1*S : Y = X1*S + Y1*C
150  PROCjoin(X1,Y1, X,Y, 3)
160  X1 = X : Y1 = Y
170 NEXT I
490 END
```

To show the speed of this algorithm add:

```
115 FOR R = 0.5 TO 5 STEP 0.5
180 NEXT R
```

to get a set of "nested" circles.

To take full advantage of the BBC micro's speed try the following changes:

```
130 FOR I% = 1 TO N
170 NEXT
```

2.9 DOODLING WITH SEMICIRCLES

In Durell's *Elementary Geometry* there are some nice examples of the kind of pleasing patterns that can be constructed with circles and semicircles. An example of one such pattern is shown in Fig. 2.15. In this case there is a

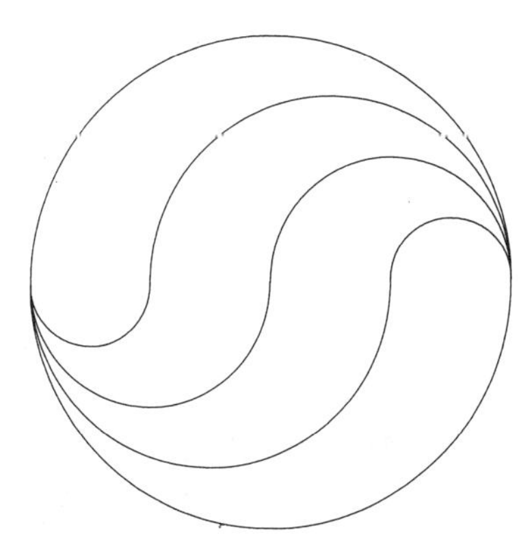

Fig. 2.15 — Durell's pattern of semi-circles.

bounding circle of radius R and centre (0,0), say. Within this circle we have sets of curves that can be constructed, from right-to-left, by drawing an upper semicircle of radius r and centre (R−r,0) followed by a lower semicircle of radius (R−r) and centre (−r,0). In the example the values of r seem to be R/4, R/2 and 3*R/4, which we can generalise to t*R for values of a parameter t from 1/4 to 3/4 in steps in 1/4.

In writing the program it would be nice to have a procedure that could draw the top or bottom part of a circle depending upon the value of some variable F. If F=1, say, then we want an upper semicircle, and if F=−1, say, we want the lower semicircle. Such a **two-state** variable is often called a **flag**. Suppose we have such a procedure then the main part of the program might look like:

```
 10 REM Prog.2.13 - Durell circle construction
 49 :
 50 MODE 1 : PROCsetup
 60 OX = 0.5*SW : OY = 0.5*SH
 70 SX = 100 : SY = 100
 99 :
100 R = 5
110 PROCsemicircle(0,0,R,1)
120 PROCsemicircle(0,0,R,-1)
130 FOR t = 1/4 TO 3/4 STEP 1/4
140  r = t*R
150  PROCsemicircle(R-r,0,r,1)
160  PROCsemicircle(-r,0,R-r,-1)
170 NEXT t
490 END
```

and we can fill in the details of the semicircle procedure definition in one of a number of ways. For example, we can adapt the "efficient circle" algorithm to give:

```
1100 DEF PROCsemicircle(XC,YC,RC,F)
1110  A = F*PI/20 : S = SIN(A) : C = COS(A)
1120  XP = RC : YP = 0
1130  FOR P = 1 TO 20
1140   XQ = XP*C - YP*S : YQ = XP*S + YP*C
1150   X1 = XQ+XC :Y1 = YQ+YC :X  = XP+XC :Y  = YP+YC
1160   PROCjoin(X1,Y1, X,Y, FC)
1170   XQ = XP : YQ = YP
1180  NEXT P
1190 ENDPROC
```

2.10 THE LADDER, ASTROID AND ELLIPSE

Returning to the ladder problem that started this chapter we can now develop some related properties of loci and envelopes. First, then, the simple envelope program:

```
 10 REM Prog.2.14 - Envelope of the Astroid
 49 :
 50 MODE 1 : PROCsetup
 60 OX = 0.5*SW : OY = 0.5*SH
 70 SX = 100 : SY = 100
 99 :
100 L = 5
110 FOR T = 0 TO 6.3 STEP PI/20
120  G = L*COS(T)
130  W = L*SIN(T)
140  PROCjoin(G,0, 0,W, 3)
150 NEXT T
490 END
```

We can easily adapt this to show both the positions of the ladder (the tangents) in red and the locus of the midpoint (circle) in white.:

```
 10 REM Prog.2.15 - Locus of the centre of the ladder
 49 :
 50 MODE 1 : PROCsetup
 60 OX = 0.5*SW : OY = 0.5*SH
 70 SX = 100 : SY = 100
 99 :
100 L = 5
110 X1 = L/2 : Y1 = 0
120 FOR T = 0 TO 6.3 STEP PI/20
130  G = L*COS(T) : W = L*SIN(T)
140  PROCjoin(G,0, 0,W, 1)
150  X = G/2 : Y = W/2
160  PROCjoin(X1,Y1, X,Y, 3)
170  X1 = X : Y1 = Y
180 NEXT T
490 END
```

As in the previous examples only small changes are necessary to trace the

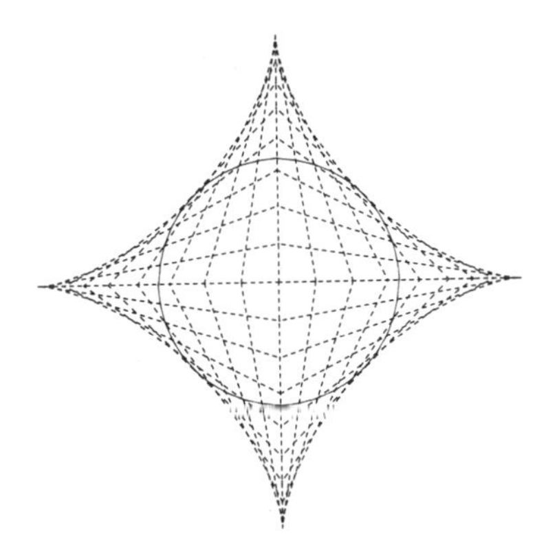

Fig. 2.16.

locus of a general point on the ladder. Try illustrating this for various values of the factor F.

If we ignore the tangents we just have a program for the elliptic locus:

```
 10 REM Prog.2.16 - Parametric equation of an ellipse
 49 :
 50 MODE 1 : PROCsetup
 60 OX = 0.5*SW : OY = 0.5*SH
 70 SX = 100 : SY = 100
 99 :
100 L = 5 : F = 1/4
110 X1 = F*L : Y1 = 0
120 FOR T = 0 TO 6.3 STEP PI/20
130  X = F*L*COS(T)
140  Y = (1-F)*L*SIN(T)
150  PROCjoin(X1,Y1, X,Y, 3)
160  X1 = X : Y1 = Y
170 NEXT T
490 END
```

Now lines 130, 140 carry the parametric equations of an ellipse. We can analyse the intersection of two neighbouring positions of the ladder given by angles T and T+dT:

Solving $y = L{*}SIN(T) - x{*}TAN(T)$
and $y = L{*}SIN(T + dT) - x{*}TAN(T + dT)$ gives

$$x = \frac{L{*}(SIN(T + dT) - SIN(T))}{TAN(T + dT) - TAN(T)} = \frac{L{*}(SIN(T + dT) - SIN(T))/dT}{(TAN(T + dT) - TAN(T))/dT}$$

If we consider the limit of this coordinate as dT tends to zero we can observe that the top becomes the definition of the derivative of L*SIN(T), which we know to be L*COS(T), and the bottom becomes the derivative of TAN(T), which we know to be sec(T) ↑ 2. Thus the limiting position of the x coordinate of the two neighbouring lines is given by:

$$x=L*COS(T)/sec(T)\uparrow 2=L*COS(T)\uparrow 3$$

and substituting this in the equation of the line to find y gives:

$$y=L*SIN(T)-L*SIN(T)*COS(T)\uparrow 2=L*SIN(T)*(1-COS(T)\uparrow 2)=L*SIN(T)\uparrow 3$$

Thus we have found the parametric equations of the enveloped curve (the astroid):

```
 10 REM Prog.2.17 - Parametric equation of an astroid
 49 :
 50 MODE 1 : PROCsetup
 60 OX = 0.5*SW : OY = 0.5*SH
 70 SX = 100 : SY = 100
 99 :
100 L = 5
110 X1 = L : Y1 = 0
120 FOR T = 0 TO 6.3 STEP PI/20
130  X = L*COS(T)^3
140  Y = L*SIN(T)^3
150  PROCjoin(X1,Y1, X,Y, 3)
160  X1 = X : Y1 = Y
170 NEXT T
490 END
```

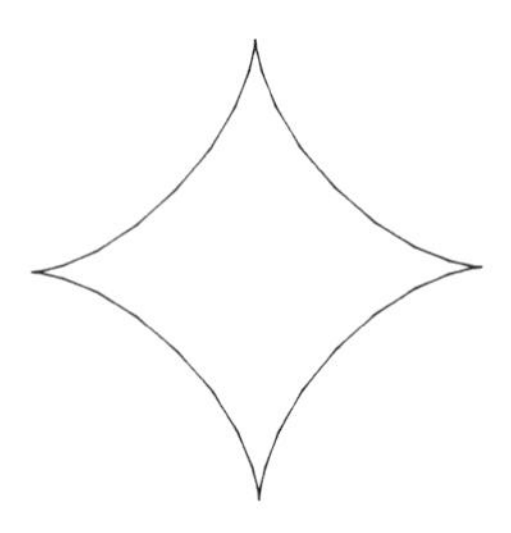

Fig. 2.17 — Locus of astroid.

We now know that the ladder at angle T is a tangent to the astroid at the point (L*COS(T)↑3,L*SIN(T)↑3) and also that a general point on the ladder is given by (F*L*COS(T), (1−F)*L*SIN(T)). If we identify these two expressions we find that this occurs when F=COS(T)↑2, so we can now draw the astroid, choose a value of the parameter T, draw in a tangent and then draw the elliptic locus of the point on the ladder given by the corresponding value of F:

```
10 REM Prog.2.18 - Astroid, tangent and ellipse
49 :
50 MODE 1 : PROCsetup
60 OX = 0.5*SW : OY = 0.5*SH
```

```
 70 SX = 100 : SY = 100
 99 :
100 L = 5
105 REM Draw the white astroid
110 X1 = L : Y1 = 0
120 FOR T = 0 TO 6.3 STEP PI/20
130  X = L*COS(T)^3
140  Y = L*SIN(T)^3
150  PROCjoin(X1,Y1, X,Y, 3)
160  X1 = X : Y1 = Y
170 NEXT T
180 :
190 REM Now draw four red tangents
200 T = PI/3 : XT = L*COS(T) : YT = L*SIN(T)
210 PROCjoin(XT,0, 0,YT, 1)
220 PROCjoin(0,YT, -XT,0, 1)
230 PROCjoin(-XT,0, 0,-YT, 1)
240 PROCjoin(0,-YT, XT,0, 1)
250 :
290 REM Now draw a yellow ellipse
300 F = COS(T)^2
310 X1 = F*L : Y1 = 0
320 FOR A = 0 TO 6.3 STEP PI/20
330  X = F*L*COS(A)
340  Y = (1-F)*L*SIN(A)
350  PROCjoin(X1,Y1, X,Y, 2)
360  X1 = X : Y1 = Y
370 NEXT A
490 END
```

Thus we have an inscribed ellipse that is tangent to the astroid at four places. As a final variation for this section we can now easily draw a family of such

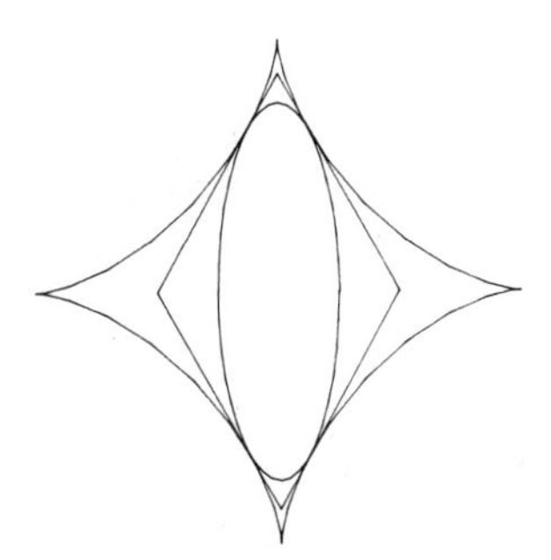

Fig. 2.18 — Astroid, tangent and ellipse.

tangential ellipses by varying the value of the parameter T in a loop:

```
 10 REM Prog.2.19 - Ellipse inscribed in an astroid
 49 :
 50 MODE 1 : PROCsetup
 60 OX = 0.5*SW : OY = 0.5*SH
 70 SX = 100 : SY = 100
 99 :
100 L = 5
105 REM Draw a red astroid
```

```
110 X1 = L : Y1 = 0
120 FOR T = 0 TO 6.3 STEP PI/20
130  X = L*COS(T)^3
140  Y = L*SIN(T)^3
150  PROCjoin(X1,Y1, X,Y, 1)
160  X1 = X : Y1 = Y
170 NEXT T
180 :
190 REM Now draw white ellipses
200 FOR T = PI/20 TO PI/2 STEP PI/20
210  F = COS(T)^2
220  X1 = F*L : Y1 = 0
230  FOR A = 0 TO 6.3 STEP PI/20
240   X = F*L*COS(A)
250   Y = (1-F)*L*SIN(A)
260   PROCjoin(X1,Y1, X,Y, 2)
270   X1 = X : Y1 = Y
280  NEXT A
290 NEXT T
490 END
```

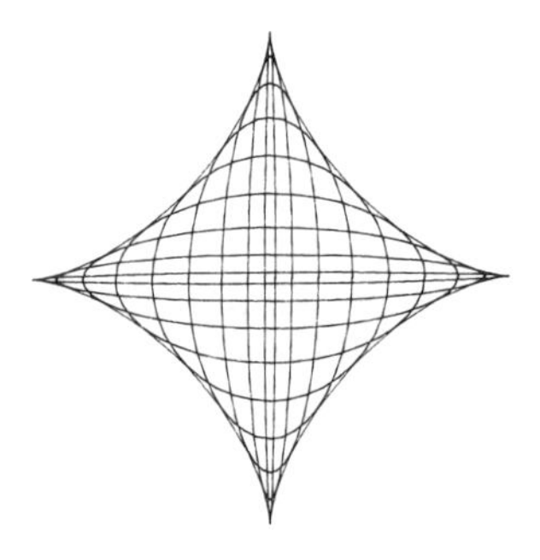

Fig. 2.19 — Family of inscribed ellipses.

2.11 GOING STRAIGHT

This chapter started with drawing lines and then used them to produce curves. In this final section we return to the subject of lines. We saw that the general position of the painter's feet on the ladder joining (G,0) to 0,W) could be represented in terms of a parameter F by ((1−F)*G*,F*W) and we can generalise this to give any point R(x,y) on the line joining P(xp,yp) to Q(xq,yq) as:

$$x=(1-t)*xp+t*xq \qquad y=(1-t)*yp+t*yq$$

Thus R is at P when t=0 and R is at Q when t=1. For values of t between 0 and 1 the point R lies between P and Q on the line-segment PQ. We can illustrate this conventional piece of theory by a small program:

```
 10 REM Prog.2.20 - The vector equation of a line
 49 :
 50 MODE 1 : PROCsetup
 60 OX = 0.5*SW : OY = 0.5*SH
 70 SX = 100 : SY = 100
 99 :
100 xp = -2 : yp = 4
110 xq = 3  : yq = 2
120 PROCjoin(0,0, xp,yp, 1)
130 PROCjoin(xp,yp, xq,yq, 1)
140 PROCjoin(xq,yq, 0,0, 1)
150 FOR t = 0.2 TO 0.8 STEP 0.2
160  x = (1-t)*xp + t*xq
170  y = (1-t)*yp + t*yq
180  PROCjoin(0,0, x,y, 3)
190 NEXT t
490 END
```

We can use this representation of a line to find the point where two lines

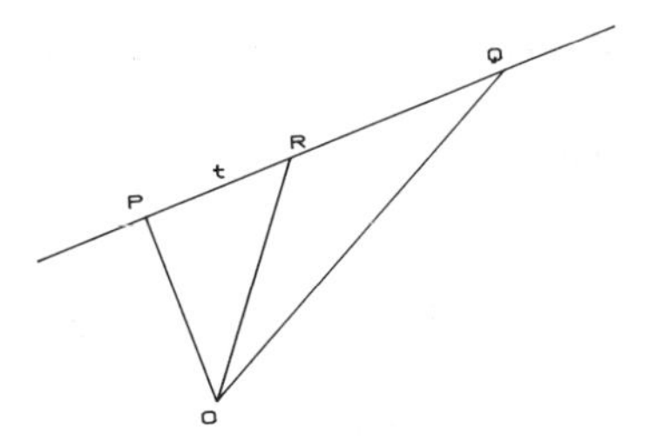

Fig. 2.20 — Vector equation of a line.

intersect. A general point (x,y) on the line joining (x1,y1) to (x2,y2) can be represented by the parameter t to give:

$$x=(1-t)*x1+t*x2 \qquad y=(1-t)*y1+t*y2$$

and a general point (x,y) on the line joining (x3,y3) to (x4,y4) can be represented by the parameter s to give:

$$x=(1-s)*x3+s*x4 \qquad y=(1-s)*y3+s*y4$$

If we solve these equation simultaneously to find the values of s and t for the point of intersection (x,y) of the two lines we have:

$$t*(x2-x1)+s*(x3-x4)=x3-x1$$
$$t*(y2-y1)+s*(y3-y4)=y3-y1$$

Now the general solution to the 2×2 set of simultaneous equations:

$$t*A+s*B=C$$
$$t*D+s*E=F$$

is given by:

$$t=(C*E-B*F)/(A*E-B*D)$$
$$s=(A*F-C*D)/(A*E-B*D)$$

so we can now find the values of t and s, and hence x and y. The following procedure does this arithmetic for us:

```
1200 DEF PROCintersect(x1,y1,x2,y2,x3,y3,x4,y4)
1210  A = x2-x1  :  B = x3-x4  :  C = x3-x1
1220  D = y2-y1  :  E = y3-y4  :  F = y3-y1
1230  den = A*E  - B*D
1240  t = (C*E  - B*F)/den
1250  s = (A*F  - C*D)/den
1260  x = (1-t)*x1  + t*x2
1270  y = (1-t)*y1  + t*y2
1290 ENDPROC
```

If the values of both t and s lie between 0 and 1 then the point of intersection is an internal point on both line segments. As an example of an application of this procedure consider generating four random points and joining them to form a closed quadrilateral:

```
 10 REM Prog.2.21  - Random quadrangles
 49 :
 50 MODE 1 : PROCsetup
 60 OX = 0.5*SW  : OY = 0.5*SH
 70 SX = 100 : SY = 100
 99 :
100 DIM x(4),y(4)
110 x(1) = 4 - 8*RND(1) : y(1) = 4 - 8*RND(1)
120 FOR i = 2 TO 4
130  x(i) = 4 - 8*RND(1) : y(i) = 4 - 8*RND(1)
140  PROCjoin(x(i-1),y(i-1), x(i),y(i), 3)
150 NEXT i
160 PROCjoin(x(4),y(4), x(1),y(1), 3)
490 END
```

It is easy enough for our eyes to detect whether, or not, the quadrilateral is "proper" or self-intersecting but it is far less apparent to the computer! See if you can use PROCintersect to solve the problem. Are there any easier tests?

As a final example for this chapter we can use the ideas from this section to draw the Pappus diagram:
In this the points P1,Q1,R1 lie on one line and the points P2,Q2,R2 on another. The point A is the intersection of the lines P1.Q2 and P2.Q1, similarly B is defined by R1.P2 and R2.P1, and C by Q1.R2 and Q2.R1. The theory is that A, B and C lie on the same line.

To turn this into a program we need to specify the positions of points as generally as possible. The positions of P1 and R1 define the first line, and the position of Q1 on this can be fixed by some parameter t. Similarly, P2 and R2 define the second line and Q2 be fixed in this by a parameter s. To draw the extended version of the line P1.Q1.R1 we could join the point given by

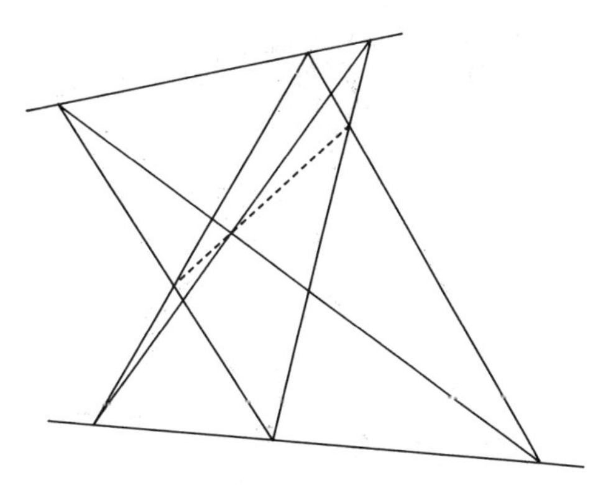

Fig. 2.21 — The Pappus diagram.

parameter value −1 to the point given by parameter value +2, and similarly for P2.Q2.R2. The six joins like P1.Q2 can be drawn unicursively (i.e. without the need to MOVE the pen between lines). We can then use PROCintersect three times to find the coordinates of A, B and C. If we draw AC we just need to verify that B lies on it:

```
  10 REM Prog.2.22   - The Pappus diagram
  49 :
  50 MODE 1 : PROCsetup
  60 OX = 0.5*SW : OY = 0.5*SH
  70 SX = 100 : SY = 100
  99 :
 100 P1x = -2 : P1y = 3 : R1x = 3 : R1y = 4
 110 t = 0.8 : Q1x = (1-t)*P1x + t*R1x : Q1y = (1-t)*P1y + t*R1y
 120 P2x = -3 : P2y = -1 : R2x = 2.5 : R2y = -3
 130 s = 0.4 : Q2x = (1-s)*P2x + s*R2x : Q2y = (1-s)*P2y + s*R2y
 140 X1 = 2*P1x - R1x : Y1 = 2*P1y - R1y
 150 X = -P1x + 2*R1x : Y = -P1y + 2*R1y
 160 PROCjoin(X1,Y1, X,Y, 3)
 170 X1 = 2*P2x - R2x : Y1 = 2*P2y - R2y
 180 X = -P2x + 2*R2x : Y = -P2y + 2*R2y
 190 PROCjoin(X1,Y1, X,Y, 3)
 200 PROCjoin(P1x,P1y, R2x,R2y, 2)
 210 PROCjoin(R2x,R2y, Q1x,Q1y, 2)
 220 PROCjoin(Q1x,Q1y, P2x,P2y, 2)
 230 PROCjoin(P2x,P2y, R1x,R1y, 2)
 240 PROCjoin(R1x,R1y, Q2x,Q2y, 2)
 250 PROCjoin(Q2x,Q2y, P1x,P1y, 2)
 260 PROCintersect(P1x,P1y,Q2x,Q2y,Q1x,Q1y,P2x,P2y)
 270 Ax = x : Ay = y
 280 PROCintersect(Q1x,Q1y,R2x,R2y,Q2x,Q2y,R1x,R1y)
 290 Cx = x : Cy = y
 300 PROCjoin(Ax,Ay, Cx,Cy, 1)
 310 PROCintersect(R1x,R1y,P2x,P2y,P1x,P1y,R2x,R2y)
 320 Bx = x : By = y
 330 PROCdot(Bx,By, 3)
 490 END
 499 :
1200 DEF PROCintersect(x1,y1,x2,y2,x3,y3,x4,y4)
1210  A = x2-x1 : B = x3-x4 : C = x3-x1
1220  D = y2-y1 : E = y3-y4 : F = y3-y1
1230  den = A*E - B*D
```

```
1240  t = (C*E - B*F)/den
1250  s = (A*F - C*D)/den
1260  x = (1-t)*x1 + t*x2
1270  y = (1-t)*y1 + t*y2
1290 ENDPROC
```

3

Curves — plain

In the last chapter we saw how a circle could be produced as a locus — the path taken by the painter's foot standing in the middle of a slipping ladder. This is only one of many different ways of defining or representing a circle. To use the computer's coordinate system it is convenient to move the origin to the middle of the screen, to fix the circle's radius as R and to let (X,Y) be the coordinates of a general point on the circle. In that system the locus approach gave these equations for X and Y:

X=R*COS(T)   Y=R*SIN(T)

The variable T is a **parameter** and both X and Y are functions of T. To specify a complete circle we need to fix a set of values for T. Since we use its sine and cosine it is reasonable to consider T as an angle. If we measure angles in degrees then a suitable range of values for T is from 0 to 360 degrees. As most computers use **radians,** rather than degrees, for measuring angles then the equivalent range in from 0 to 2*PI radians.

The following section shows how this particular example can lead to a general approach for illustrating curves defined by such **parametric equations.**

3.1 PARAMETRIC FORM

In order to plot a continuous curve we need to store the coordinates of the first point (X1,Y1) on the curve, compute successive points (X,Y) on the cuirve, **join** (X1,Y1) to (X,Y) and **archive** (X,Y) in (X1,Y1). To compute successive points we can use a FOR-NEXT loop with T as the variable.

For the circle we want T to range from 0 to 2*PI, which is a shade less than 6.3, and we just need to choose a suitable step for incrementing T. The smaller the step we use the more points will be computed and so the curve should be represented more accurately, but also drawn more slowly. Conversely, with a bigger step we compute fewer points, and obtain a cruder approximation but drawn faster.

Because of the finite resolution of the screen there is a limit to the accuracy achievable and so it will often be found that around 50 points gives a reasonable compromise between speed and accuracy. In our example, then, a step of 0.1 for T seems reasonable.

```
 10 REM Prog.3.1 - A parametric curve plotter
 49 :
 50 MODE 1 : PROCsetup
 60 OX = 0.5*SW : OY = 0.5*SH
 70 SX = 100 : SY = 100
 99 :
100 R = 4
110 X1 = R : Y1 = 0
120 FOR T = 0 TO 6.3 STEP PI/18
130  X = R*COS(T)
140  Y = R*SIN(T)
150  PROCjoin(X1,Y1, X,Y, 3)
160  X1 = X : Y1 = Y
170 NEXT T
490 END
499 :
500 DEF PROCsetup
510  SW = 1280 : SH = 1024
520  NC = 3 : FC = 3
590 ENDPROC
599 :
600 DEF PROCdot(X,Y,PC)
610  GCOL 0,PC
620  PLOT 69, OX + X*SX, OY + Y*SY
690 ENDPROC
699 :
700 DEF PROCjoin(X1,Y1,X,Y,PC)
710  GCOL 0,PC
720  MOVE OX + X1*SX, OY + Y1*SY
730  DRAW OX + X*SX, OY + Y*SY
790 ENDPROC
799 :
```

The conversion of the procedures for other micros is covered in section 1.7.

Now we have the basic form we can easily start to 'tweak some knobs'. In order to explore different parametric forms we now only need to change lines 130 and 140. Any constants that may be needed can be defined in line 100. If a different range of parameter is wanted then we just need to change line 120. The whole routine in lines 110–170 draws just one of a family of parametric curves specified by the values of the constant(s) in line 100. Thus it could easily be enclosed in further FOR-NEXT loops to draw several members of the family, or made into a subroutine or procedure to be called by other programs.

For example we have seen that we can change the basic program to draw an ellipse by the modifications:

```
110 A = 400 : B = 200
130  X = A*COS(T)
140  Y = B*SIN(T)
```

and obtain a satisfying pattern of ellipses shown in Fig. 3.1 by adding:

```
110 FOR A = 50 TO 500 STEP 50
115  B = 25000/A
180 NEXT A
```

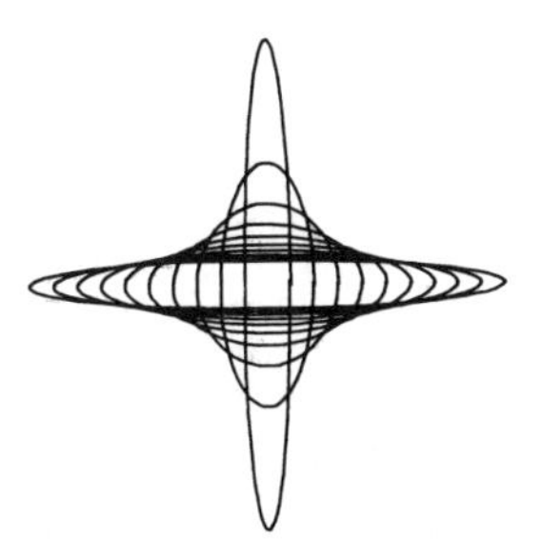

Fig. 3.1

The corresponding changes for other versions are almost identical.

Similarly we can obtain the well-known **Lissajous** figures by slight changes to the parametric forms in lines 130 and 140. These curves are given by using SIN and COS functions of different periods — this can be achieved by:

```
130  X = A*COS(M*T)
140  Y = B*SIN(N*T)
```

for different values of M and N:

```
 10 REM Prog.3.2 - Lissajous figures
 49 :
 50 MODE 1 : PROCsetup
 60 OX = 0.5*SW : OY = 0.5*SH
 70 SX = 100 : SY = 100
 99 :
100 A = 4 : B = 3
105 M = 1 : N = 4
110 X1 = R : Y1 = 0
120 FOR T = 0 TO 6.3 STEP PI/18
130  X = A*COS(M*T)
140  Y = B*SIN(M*T)
150  PROCjoin(X1,Y1, X,Y, 3)
160  X1 = X : Y1 = Y
170 NEXT T
490 END
```

Figure 3.2 shows some curves produced from this program. The first one uses M=1 and N=4, try to find which pairs of values of M and N produced the others. See if you detect any patterns. You may need to change the step

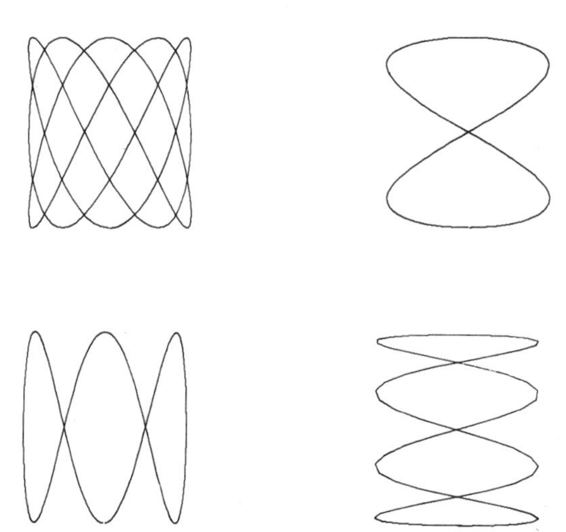

Fig. 3.2

size for T in line 120 for some of the more complex curves.

The **astroid**, which we met in the previous chapter, has the parametric equations:

X=A*(COS(T)) ↑ 3   Y=B*(SIN(T)) ↑ 3

However, there can be trouble in using the exponential operator "↑" ("raising to the power") with negative values. This problem can be side-stepped by avoiding the ↑ operator completely:

```
125  C = COS(T) : S = SIN(T)
130  X = A*C*C*C
140  Y = B*S*S*S
```

In fact, some quite extraordinary effects can be achieved by trying variations on lines 130 and 140. Figure 3.3 shows some examples. The first one used:

X=A*C*(2−S)*S   Y=B*S*(1−C)

Try finding some interesting examples of your own.

To cope with the general parametric form:

X=A*(COS(T)) ↑ M   Y=B*(SIN(T)) ↑ N

we must handle the signs to ensure that we do not take powers of negative values. (The reason for this is that most versions of BASIC perform the "↑" operation by taking logarithms, which fails for negative values.) The following dodge uses the BASIC functions SGN and ABS. SGN(X) has a

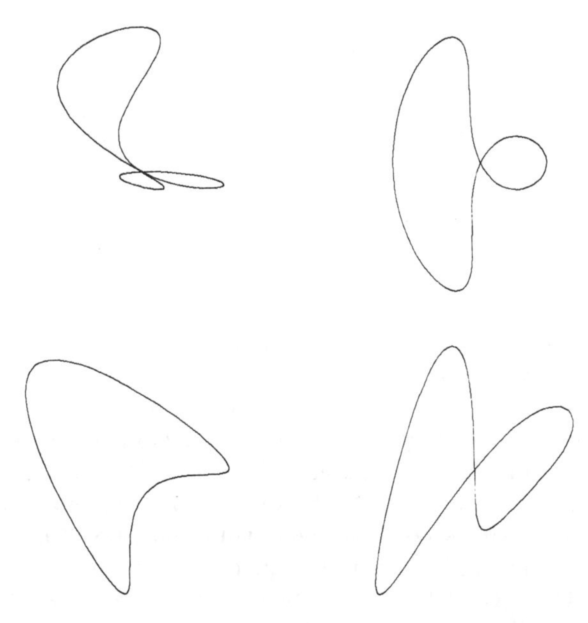

Fig. 3.3

value of −1, 0 or 1 depending whether X is negative, zero or positive. ABS(X) takes the absolute value (or modulus) of X, i.e. it ignores any minus signs. If the power to which the number is raised is even then the result is positive. If it is odd then the sign of the result is the same as the sign of the original number. A number is odd if it leaves a remainder of two on division by 2 and a convenient way of performing this is provided by the Basic operator X MOD Y, which produces the remainder on dividing X by Y:

```
130  X = A*(ABS(C))^M
135  IF (M MOD 2)=1 THEN X = X*SGN(C)
140  Y = B*(ABS(S))^N
145  IF (N MOD 2)=1 THEN Y = Y*SGN(S)
```

Unfortunately the MOD operator is not common to all versions of Basic. An alternative test for odd and even numbers uses the function INT, which just returns the integer part of a number. Thus a number is even if, after division by 2, the result is a whole number. In some cases then we need the following versions of lines 135 and 145:

```
135 : IF (M/2) > INT(M/2) THEN LET X = X*SGN(C)
145 : IF (N/2) > INT(N/2) THEN LET Y = Y*SGN(S)
```

Figure 3.4 shows the curve produced by setting M=1 and N=5. This family of curves has been called the "lips" curves. Try exploring the effects of different values of M and N. What happens if either M or N or both tend to infinity? See if you can extend the definition in such a way that M and N need not be integers. What happens if either M or N or both tend to 0?

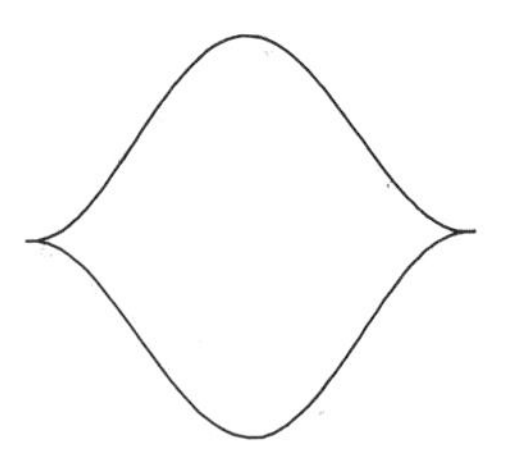

Fig. 3.4

So far we have concentrated on parametric equations involving the rather well-behaved trigonometric functions: sine and cosine. These are well-behaved since they can be scaled to fit entirely on the screen and they do not suddenly shoot off to infinity. In adapting the parametric program for less well-behaved equations we may run into two main classes of problem.

The first involves either X or Y becoming infinite by, perhaps, a division by zero. This kind of problem will usually result in the program stopping with some kind of error message on the screen.

The second kind of problem occurs when the first problem has been 'side-stepped'. In this case, for example, one point on the curve might have a very large positive Y coordinate and the next point used might have a very large negative Y coordinate. Presumably between these two points which we have sampled there must have been a singularity at which Y goes to infinity. If we join these two points with a straight line it will certainly not look like a small segment of the curve.

As an example consider the curve produced by parametric equations:

$$X=A*\sec(T) \qquad Y=B*TAN(T)$$

Now TAN is a Basic function whereas "sec" is not. The secant, sec(T), of the angle T is the reciprocal of the cosine of T and so we can use the equivalent form:

$$X=A/COS(T) \qquad Y=B*TAN(T)$$

This parametric form should not be well-behaved for values of T between 0 and 2*PI since X will become infinite when COS(T)=0, which happens at T=PI/2 and T=3*PI/2, and Y will become infinite when TAN(T) becomes infinite, which also happens at the same values of T=PI/2 and T=3*PI/2.

Making the following changes to Prog. 3.1:

```
110 A = 3 : B = 3
130  X = A/COS(T)
140  Y = B*TAN(T)
```

produces the display shown in Fig. 3.5 and the program does not "crash" — why?

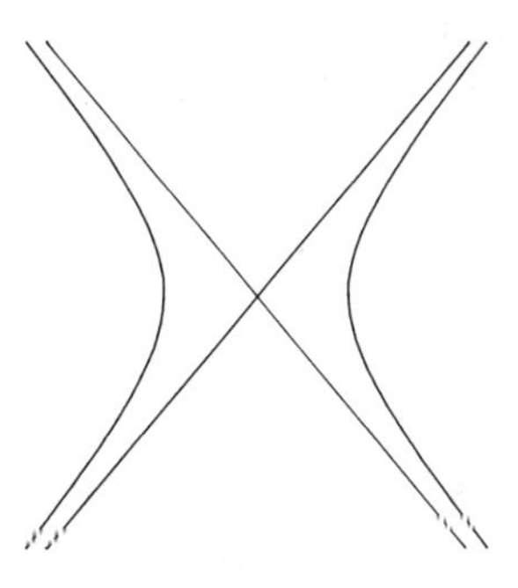

Fig. 3.5

We have been 'lucky' in dodging the singularities at T=PI/2 and T=3*PI/2 by the choice of the staring value and the steps for T in the FOR-NEXT loop. Now PI/2 is about 1.571 and our values of T, starting from 0 in steps of 0.1, straddle this with values of 1.5 and 1.6. The coordinates of the point corresponding to T=1.5 are roughly (4200,4200) and those for T=1.6 are (−10000,−10000). These are certainly not points physically close to each other on the curve, and hence the line joining them should not really be allowed to appear. In fact, the curve in question is a hyperbola and so the lines look like the asymptotes, but we cannot really claim that we set out to draw them! The problem is even more acute with a microcomputer like the ZX-Spectrum which will not allow points to be plotted nor lines drawn if they go off the screen area. In this case, then, we need to use a "trap" to ensure that the only lines drawn are ones which are wholly within the display screen area.

A neat way to do this is to use the notion of "logical" operators and variables that BASIC provides. The operators >,< and = result in a value that is either true or false. These are usually represented within the computer by the numbers −1 (or sometimes 1) and 0. You can test this by trying the following two direct commands:

```
PRINT 1+1=2      PRINT 1+1=3
```

Such logical operations can be connected by the additional operators AND, OR and NOT — try the following:

```
PRINT (4>3) AND (4>5)
PRINT (4>3) OR (4>5)
PRINT (4>3) AND NOT (4>5)
```

Using this idea we can define variables, which we will call OK1 and OK, to take the value **true** if points (X1,Y1) or (X,Y) appear on the screen and

false otherwise. If both OK1 and OK are **true** then we can safely join the two points with a line. This idea is incorporated in the following program:

```
 10 REM Prog.3.3 - A safer parametric curve plotter
 49 :
 50 MODE 1 : PROCsetup
 60 OX = 0.5*SW : OY = 0.5*SH
 70 SX = 100 : SY = 100
 80 MX = 5 : MY = 5
 99 :
100 A = 3 : B = 4
110 X1 = A : Y1 = 0
115 OK1 = (ABS(X1)<MX) AND (ABS(Y1)<MY)
120 FOR T = 0 TO 6.3 STEP PI/18
130  X = A/COS(T)
140  Y = B*TAN(T)
145  OK = (ABS(X)<MX) AND (ABS(Y)<MY)
150  IF (OK1 AND OK) THEN PROCjoin(X1,Y1, X,Y, 3)
160  X1 = X : Y1 = Y : OK1 = OK
170 NEXT T
490 END
```

We may not always be so lucky as to accidentally dodge the "nasty bits" of a curve. If we change line 130 to:

```
130 X=A/SIN(T)
```

then we invite trouble straightaway, since there is a singularity at T=0. There are some sophisticated dodges to anticipate when the computer is going to produce an error message, using the ON ERROR... statement, but this is not provided in all micros. The obvious dodge, then, is to adjust the range of values for T slightly so that any singularities are side-stepped. If we start T at, say, 0.01 instead of 0 then the program should not "crash". In the programs of this section we will need to alter just lines 110 and 120 to do this. Figure 3.6 shows the outcome.

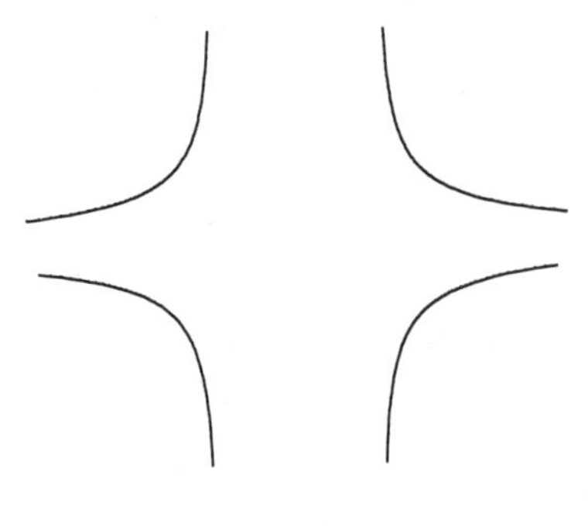

Fig. 3.6

The last thing that remains is to invite the reader to have fun exploring the kinds of curves that can result from experimenting with the use of different functions of T in lines 130 and 140. As well as the trigonometric functions SIN, COS and TAN there are other standard functions in BASIC that we have not explored, such as EXP, LOG (or LN) as well as other functions such as ABS, INT, SGN, SQR and RND that are not usually met in coordinate geometry books! Using these together with the arithmetic operators +, −, *, /, ↑, brackets () and the logical operators >, >=, <, <=, = <> together with AND, OR and NOT there is an enormous scope for the imagination to roam. Figure 3.7 was produced by the following

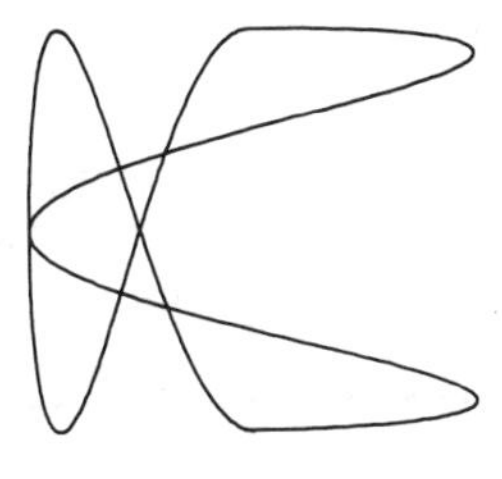

Fig. 3.7

changes to lines 130 and 140 that show how "piecewise" definitions of curves can also be included:

```
130  X = A*( SIN(T)*(T < PI) + SIN(3*T)*(T >= PI))
140  Y = B*( COS(3*T)*(T < PI) + COS(T)*(T >= PI))
```

For the sake of convenience some "standard" parametric forms have been collected together here, but many geometry books (particularly Lockwood) will contain others. You may find it necessary to think a little about an appropriate parameter range for some of these curves. We have concentrated on a range from 0 to 2*PI, but some examples might need T to take all real values.

X=COS(2*T)+COS(T)   Y=SIN(2*T)+2*SIN(T)
X=3*COS(T)−COS(3*T)   Y=3*SIN(T)−SIN(3*T)
X=2/COS(T)   Y=3*TAN(T)
X=T↑2/(1+T↑2)   Y=T↑3/(1+T↑2)
X=T↑2   Y=2*T
X=(T↑2−1)/(T↑2+1)   Y=T*(T↑2−1)/(T↑2+1)
X=T   Y=1/T
X=COS(T)↑3   Y=SIN(T)↑3
X=3*COS(T)   Y=2*SIN(T)
X=(1−T↑2)/(1+T↑2)   Y=2*T/(1+T↑2)
X=3*T↑2   Y=2*T↑3
X=SIN(T)↑2   Y=COS(T)↑2
X=2*T   Y=3*T

3.2 CARTESIAN FORM

In the last section we used the general parametric form in which both X and Y were functions of a parameter T:

$$X=f(T) \qquad Y=g(T)$$

A special case of this occurs when X and T are identical, in which case Y can be considered as a function of X:

$$Y=F(X)$$

Thus the programs of the last section can be very easily adapted to plot cartesian forms such as Y=SIN(X) although, in this case, we will now have to fix the range of values that the independent variable X can take:

```
 10 REM Prog.3.4 - A cartesian curve plotter
 49 :
 50 MODE 1 : PROCsetup
 60 OX = 0.5*SW : OY = 0.5*SH
 70 SX = 100 : SY = 100
 99 :
100 DEF FNY(X) = SIN(X)
110 XL = -PI : XH = PI : XS = (XH-XL)/32
120 X1 = XL : Y1 = FNY(XL)
130 FOR X = XL TO XH STEP XS
140  Y = FNY(X)
150  PROCjoin(X1,Y1, X,Y, 3)
160  X1 = X : Y1 = Y
170 NEXT X
490 END
```

In this program the origin has again been shifted to the centre of the screen.

Fig. 3.8

To change the function plotted we have only to alter the function definition in line 100, and to change the values of X for which the curve is plotted we have only to change the values XL and XH in line 110.

The techniques of Prog. 3.3 can also be easily applied to this program to prevent the program from "crashing" if points appear outside the screen area.

In fact, the curve drawing routine between lines 110 to 170 could easily be rewritten as a subroutine or procedure to allow several curves to be plotted, perhaps using different colours. A BBC program to illustrate this is:

```
   10 REM Prog.3.5 - plotting several cartesian curves
   49 :
   50 MODE 1 : PROCsetup
   60 OX = 0.5*SW : OY = 0.5*SH
   70 SX = 100 : SY = 100
   99 :
  100 Y$ = "SIN(X)"
  110 XL = -PI : XH = PI : XS = (XH-XL)/32
  120 PROCplot(Y$, 3)
  130 Y$ = "COS(X)"
  140 PROCplot(Y$, 2)
  490 END
  499 :
10100 DEF PROCplot(Y$, PC)
10110  X1 = XL : X = X1 : Y1 = EVAL(Y$)
10120  FOR X = XL TO XH STEP XS
10130   Y = EVAL(Y$)
10140   PROCjoin(X1,Y1, X,Y, PC)
10150   X1 = X : Y1 = Y
10160  NEXT X
10190 ENDPROC
```

Since there are difficulties with changing the definition of a function within a program we have made use of the EVAL function, which is unfortunately not available in all versions of Basic.

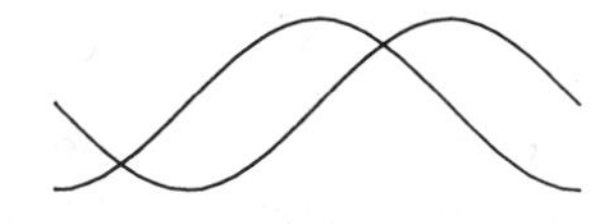

Fig. 3.9

This plots a sine graph in yellow and then a cosine graph in red.

So far the programs in this section have illustrated the plotting of curves of the form Y=F(X) in which Y is an **explicit** function of X. Obviously, using the techniques introduced at the end of the previous section, we can explore some pretty exotic functions of X using the functions and operators available to us in Basic.

However, most of the "standard" curves studied in geometry tend not to lend themselves to simple explicit forms but rather specify relations between the X- and Y-coordinates of points (X,Y) that lie on the curve. For example, the catesian equation of a circle of radius R with its centre at the origin is:

X↑2+Y↑2=R↑2

This can be rearranged to give an explicit equation for Y:

Y=SQR(R↑2−X↑2)

The changes to Prog. 3.4 to plot this are:

```
100 DEF FNY(X) = SQR(R*R-X*X)
105 R = 4
110 XL = -R : XH = R : XS = (XH-XL)/32
```

The program only draws the top half of the curve since the SQR function

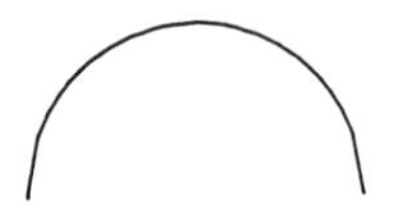

Fig. 3.10

in Basic only returns the positive square root. To obtain both halves of the curve we can first plot the function using the positive square root and then plot using the negative values. To do this we can make use of Prog. 32.5 or repeat the plotting loop twice:

```
180 FOR X = XH TO XL STEP -XS
190   Y = -FNY(X)
200   PROCjoin(X1,Y1, X,Y, 3)
210   X1 = X : Y1 = Y
220 NEXT X
```

Of course, there are snags with this kind of representation, particularly in the loss of symmetry. In general, then, if a curve, such as a circle, ellipse or astroid, has a known parametric form it is preferable to plot it using the methods of section 3.1. If it has a cartesian form that can be rearranged to give Y as an explicit function of X then the techniques of this section will enable you to plot it.

If, as is frequently the case, the **implicit** relationship that connects X and Y cannot be easily rearranged to give Y as a function of X then for each value of X in the domain an equation will have to be solved (usually be means of some numerical technique) for the corresponding possible values of Y. Such equations may produce no solutions, unique solutions or multiple solutions. On the whole such programs are both messy and slow but we will explore one approach in a later section.

3.3 POLAR FORMS

The familiar (X,Y) coordinates (the cartesian coordinates) represent just one way of fixing a point in two dimensions. Another common way is to choose a fixed point O, the origin, and a line running through it, usually thought of as pointing "East". Then any point P in the plane can be specified by an angle T, the amount of turn anti-clockwise from the fixed line, and a distance R, the length of OP. These two numbers (R,T) form the **polar coordinates** of P. From Fig. 3.11 it is easy to see the relation between the

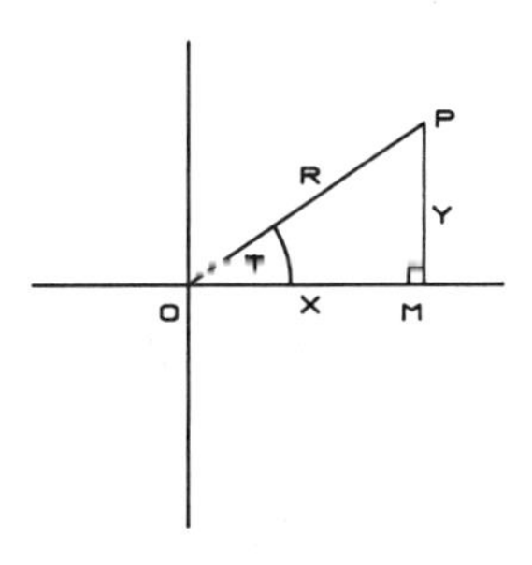

Fig. 3.11

polar coordinates (R,T) and the cartesian coordinates (X,Y) from the right-angled triangle OPM:

X=R*COS(T) Y=R*SIN(T)

A curve is usually specified in this system by defining the radial length R to be a function of the angle T, R=F(T). For example, one standard curve has the form:

R=A*(1−COS(T))

where A is some suitable constant. As T ranges from 0 to PI/2 to PI to 3*PI/2 to 2*PI so R increases from 0 to A to 2*A and then decreases to A and back to 0. To plot the curve on most micros we need to convert the polar coordinates (R,T) into cartesian coordinates (X,Y). Since R is a function of T then so are X and Y. Thus the program to plot this kind of curve is almost identical to that of Prog. 3.1:

```
 10 REM Prog.3.6 - A polar curve plotter
 49 :
 50 MODE 1 : PROCsetup
 60 OX = 0.5*SW : OY = 0.5*SH
 70 SX = 100 : SY = 100
 99 :
100 A = 2
110 X1 = 0 : Y1 = 0
120 FOR T = 0 TO 6.3 STEP 0.1
130   R = A*(1-COS(T))
140   X = R*COS(T) : Y = R*SIN(T)
150   PROCjoin(X1,Y1, X,Y, 3)
160   X1 = X : Y1 = Y
170 NEXT T
490 END
```

The curve produced from this program is shown in Fig. 3.12. This curve is

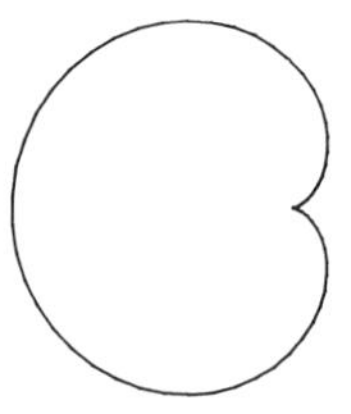

Fig. 3.12

known as the Cardioid because of its "heart-shape".

As in the previous section we now have all the equipment necessary to enable us to explore a vast range of curves just by changing the function used in line 130. For example, the Cardioid is a special case of a family of curves known as Limaçons, from a French word for a "snail". To explore the shapes of other curves in this family just change line 130 to:

```
130  LET R = A*(1 - L*COS(T))
```

and try varying the value of L. The simplest way of doing this is to add the line:

```
105 INPUT L
```

Figure 3.13 shows some curves of this family for different values of L.

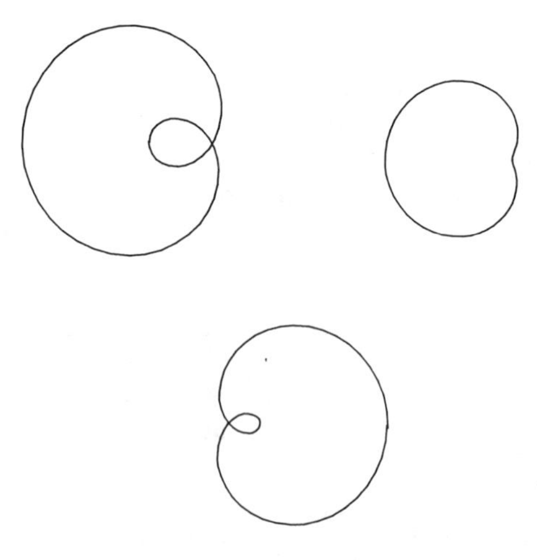

Fig. 3.13

The first used L=2, see if you can find out the other values of L used for the remaining curves.

Another very simple change to the function used in line 130 is to take a multiple of T as the angle:

```
130   R = A*(1 - L*COS(M*T))
```

and change line 105 to:

```
105 INPUT L : INPUT M
```

For some of the "fancier" curves it may be necessary to take smaller steps in the T-loop.

This simple trick gives the curves of Fig. 3.14. The values of L used were

Fig. 3.14

0.1, 1, 1.5 and 2. The values for M were 2, 3, and 5. Can you see which L went with which M to give each of the curves shown?

A further simple variant is to add a spiral effect by multiplying R by T:

```
130   R = T*A*(1 - L*COS(M*T))
```

using this with L=1 and M=3 produces Fig. 3.15:

As a final example of yet another variation on our simple theme we can experiment with raising the cosine of the angle used in line 130 to some power N:

```
130   R = A*(1 - L*COS(M*T)^N)
```

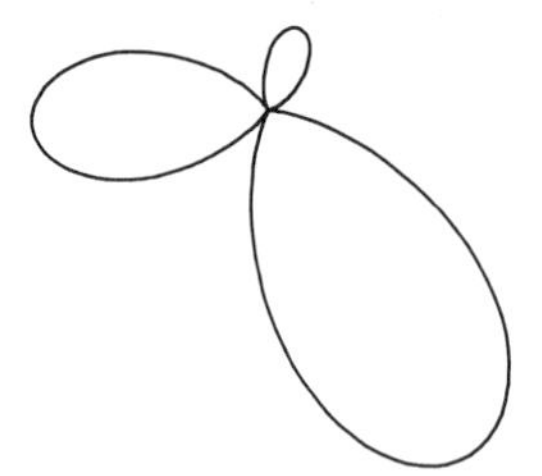

Fig. 3.15

However, since the cosine will become negative, we can anticipate some trouble with using the "raising to the power" operator ↑ . We have, though, already met a dodge to avoid this by using the SGN and ABS functions. The following amendments will work in most cases:

```
124  C = COS(M*T) : S = SGN(C) : I = ABS(C)
126  IF N=2*INT(N/2) THEN S = 1
130  R = A*(1 - L*S*I^N)
```

If we use this together with a line to input a value for N:

```
107 INPUT N
```

we can explore yet another class of curves by varying the three parameters L, M and N. The first curve in Fig. 3.16 uses L=1, M=7 and N=5 and the

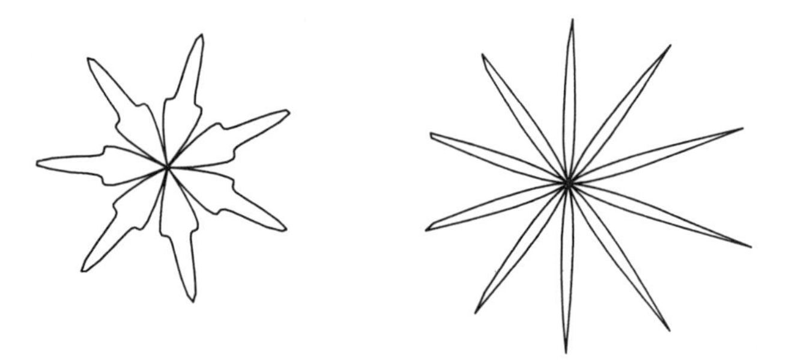

Fig. 3.16

second uses L=1, M=5 and N=0.5 — explore what happens for very large or very small N.

If you like the "theme and variations" approach you might try exploring "blends" of cosines and sines:

e.g. R=A∗COS(M∗T)+B∗SIN(N∗T)

Figure 3.17 was obtained from:

```
R=COS(4*T)+SIN(7*T)+3
```

Fig. 3.17

3.4 BIPOLAR COORDINATES

One standard construction for an ellipse uses a piece of string and two drawing-pins. The ends of the string are tied to the pins and the pins are pushed into a drawing board so that the string is quite slack. If you then put a pencil inside the string and stretch the string tight the pencil can be made to describe an arc of an ellipse as it moves from side to side — see Fig. 3.18.

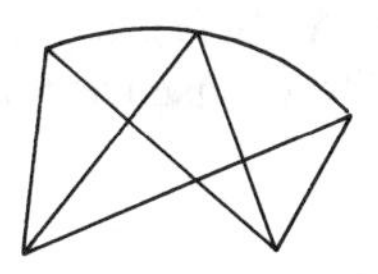

Fig. 3.18

If we take the fixed points E and F to be one unit apart and the length L of the string to be greater than 1, say L=2, then the distances R=EP and S=FP from the fixed points (the foci) are connected by:

```
R+S=L
```

Thus the point P is fixed by the two lengths R and S. If we let R vary then S can be expressed as a function of R (S=L−R) and we can investigate the locus of P as R varies.

In order to show this locus on the computer's display screen we have once more to convert from this coordinate system to the (X,Y) coordinate system of the screen. In the triangle EPF we know the three sides: EP=R, PF=S, EF=1 and so we can use the cosine rule to find the angle T at E.

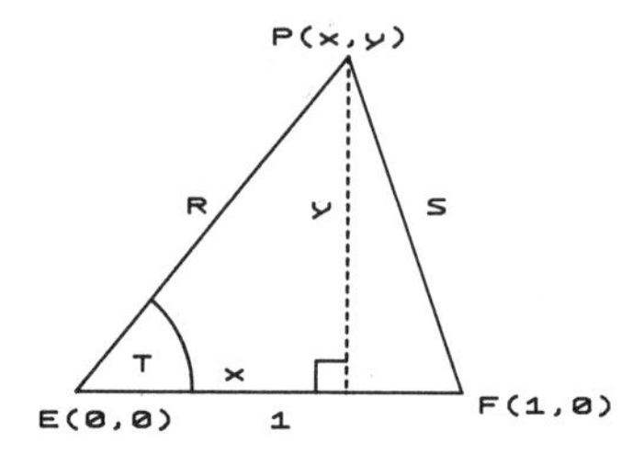

Fig. 3.19

S*S=R*R+1*1−2*R*1*COS(T)

rearranges to give:

COS(T)=(R*R+1−S*S)/(2*R)

Some values of R and S may result in "impossible" triangles but we can let the program sort that out. By putting:

D=(R*R+1−S*S)/(2*R)

we can test to see if ABS(D) is greater than 1 — which would signal an impossible triangle. Otherwise we could then use the inverse cosine function ACS to calculate T:

T=ACS(D)

We now have the polar coordinates (R,T) of P and so we can convert them to cartesian form by: X=R*COS(T), Y=R*SIN(T). There is some labour-saving available, though, since we have already calculated COS(T) as D, thus we need:

X=R*D Y=R*SQR(1−D*D)

This can now be contained in a FOR-NEXT loop using R as the variable. We must take care to avoid using R=0 because of the division required in calculating D. So we can just try an arbitrary range of values of R to see what happens.

All that remains is to fix where the origin E is to go and to scale up the lengths by some factor M to fit the screen. For display purposes we might choose to plot the different positions of the triangle EPF or to plot the locus of P. By symmetry for each position of P above the line EF there will be a corresponding position P′ given by the reflection of P in EF. A pleasing display is formed by drawing different positions of the "kite" EPFP′.

```
  10 REM Prog.3.7 - Bipolar coordinates
  49 :
  50 MODE 1 : PROCsetup
  60 OX = 0.5*SW : OY = 0.5*SH
  70 SX = 300 : SY = 300
  99 :
 100 FOR R = 0.1 TO 3 STEP 0.1
 110  S = 2 - R
 120  D = (R*R+1-S*S)/(2*R)
 130  IF ABS(D)<=1 THEN PROCkite
 140 NEXT R
 490 END
 499 :
1200 DEF PROCkite
1210  X = R*D : Y = R*SQR(1-D*D)
1220  PROCjoin(0,0, X,Y, FC)
1230  PROCjoin(X,Y, 1,0, FC)
1240  PROCjoin(1,0, X,-Y, FC)
1250  PROCjoin(X,-Y, 0,0, FC)
1290 ENDPROC
```

The output from this program is shown in Fig. 3.20:

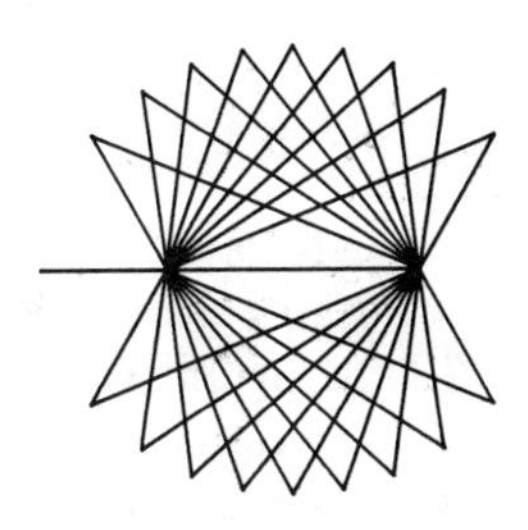

Fig. 3.20

Try adapting the program to draw the locus.

3.5 CURVES IN TIME

In many physical situations the position of a moving object is a function of a very important parameter: **time.** This section shows how two common forms of movement can be modelled with the micro. The first example is that of a bouncing ball — or, more technically, the trajectory of a projectile. The second example is that of a hound chasing a fox, which gives rise to examples of "curves of pursuit".

For a projectile launched with initial velocity U at an angle A to the horizontal and moving under the action of gravity G we can consider its motion made up from a steady horizontal velocity H=U*COS(A) (assuming a vacuum!) and motion vertically with constant acceleration −G and initial velocity V=U*SIN(A). This trajectory is, in fact, a parabola and using the equations for motion with constant acceleration we have the following expressions for the coordinates (X,Y) of the projectile P at time T after its release:

$$X=H*T \qquad Y=V*T-G*T*T/2$$

Thus X and Y are both functions of the parameter T and so the methods

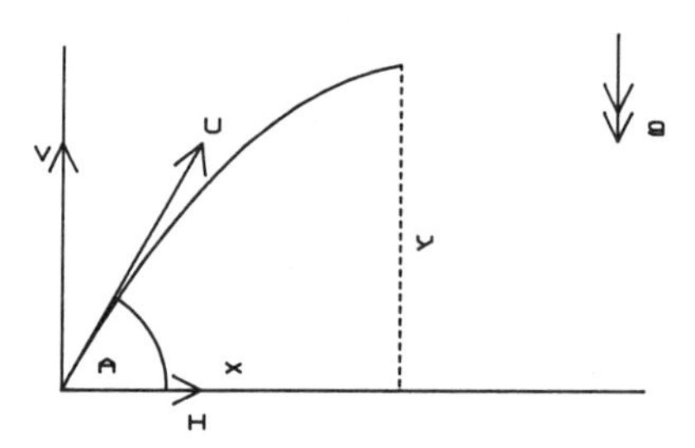

Fig. 3.21

of section 3.1 can be used to plot the trajectory. However, we may need to adopt a different approach to the programming structure that allows T to vary. In section 3.1 we used a **definite** loop (starting with: FOR T =... and finishing with: NEXT T). In this case we might wish to stop plotting the trajectory when some condition is met, like hitting the floor! We can use an **indefinite** loop which is terminated by testing to see if, for example, Y has become negative (which means the ball is buried in the ground). In BBC Basic (as in some other programming languages) there is a loop structure which starts with the command: REPEAT and finishes with the condition to be tested, in this case: UNTIL Y<=0.

Since the loop is now controlled by some condition on Y we must ensure that theparameter T is properly incremented. If T holds the current time then we can use a constant increment DT to "update the clock" by putting: T=T+DT within the loop.

```
 10 REM Prog.3.8 - Motion of a projectile
 49 :
 50 MODE 1 : PROCsetup
 60 OX = 0.1*SW : OY = 0.2*SH
 70 SX = 40 : SY = 40
 99 :
100 PROCjoin(0,0, 20,0, 1)
110 A = PI/3 : U = 10
120 H = U*COS(A) : V = U*SIN(A) : G = 9.8
130 DT = 1/16 : T = DT
```

```
140 X1 = 0 : Y1 = 0
150 REPEAT
160  X = H*T : Y = T*(V-G*T/2) : T = T+DT
170  PROCjoin(X1,Y1, X,Y, 3)
180  X1 = X : Y1 = Y
190 UNTIL Y<=0
490 END
```

The versions for IBM PC and Apple Mac will use the WHILE-WEND alternative:

```
150 WHILE Y>0
160  X = H*T : Y = T*(V-G*T/2) : T = T+DT
170  GOSUB 700 : REM PROCjoin(X1,Y1, X,Y, 3)
180  X1 = X : Y1 = Y
190 WEND
```

The ZX-Spectrum version of Basic, in common with many other micros, does not have the REPEAT-UNTIL structure. In this case the loop will finish with a test such as: IF Y>0 THEN... which will usually be followed with the much criticised GO TO statement to send you back to the start of the loop. The following version shows how we can translate the REPEAT-UNTIL structure and still convey some feeling about what is going on:

```
150 REM REPEAT
160 : LET X = H*T : LET Y = T*(V-G*T/2) : LET T = T+DT
170 : GO SUB 700 : REM PROCjoin(X1,Y1, X,Y, 3)
180 : LET X1 = X : LET Y1 = Y
190 IF Y>0 THEN GO TO 150
```

The main loop of this program could easily be surrounded by a FOR-NEXT loop for the angle A, say, to show lots of possible trajectories.

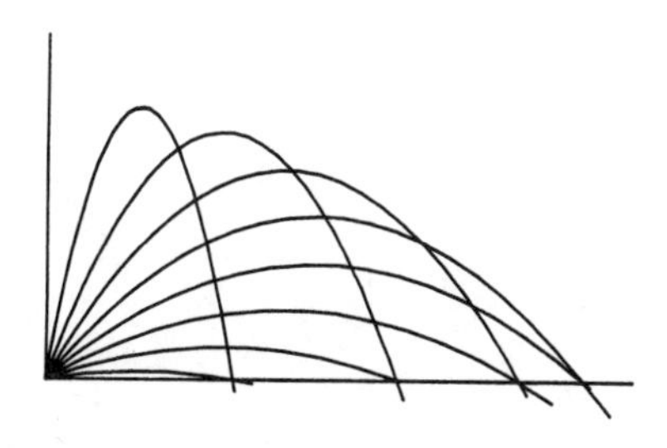

Fig. 3.22

Adapt the program to make the "ball" bounce several times. Try to introduce an "elastic constant" so that the bounces decay.

Much fun can be had with projectiles, and with other equations of motion. For example, see if you can erect some vertical "walls" and make the ball bounce horizontally as well as vertically. Later we shall see ways of displaying trajectories in three dimensions.

See if you can adopt the method to model the swing of a simple pendulum. If you feel really ambitious you could try to model the path of the pendulum bob for a hefty "swing" which takes it past the horizontal and in which the string goes slack for a period.

The second example of curves with time as the parameter is a curve of pursuit. Suppose a **fox** is running along a straight line with a steady speed and suddenly comes into the view of a **hound**. The hound, if it is inexperienced enough, will start to run towards the place where it sees the fox. But while the hound's heading in that direction the fox will have moved, and so the hound will have to change its direction, and so on...

To model this we need to specify the current coordinates (XF,YF) of the fox and (XH,YH) of the hound. We need to know the speeds of each, which will be assumed to stay constant. Let VF be the speed of the fox and VH that of the hound. We can also fix the angle AF that the path of the fox makes with the screen's horizontal. If the variable T holds the current time we must also introduce an increment DT for the intervals after which the hound "updates" its direction of pursuit.

As an illustration of "tidy" programming the following program uses a **procedure** for performing the necessary drawing in each time interval DT. We need to compute the horizontal and vertical differences, DX and DY, between the fox and the hound at time T. Using Pythagoras with these we can find the distance L between them. Now in a time interval DT the hound will move a distance DT*VH so we can introduce the ratio M=DT*VH/L between the "big" triangle representing the distances between the animals and the "little" triangle representing the hound's motion in the time interval DT.

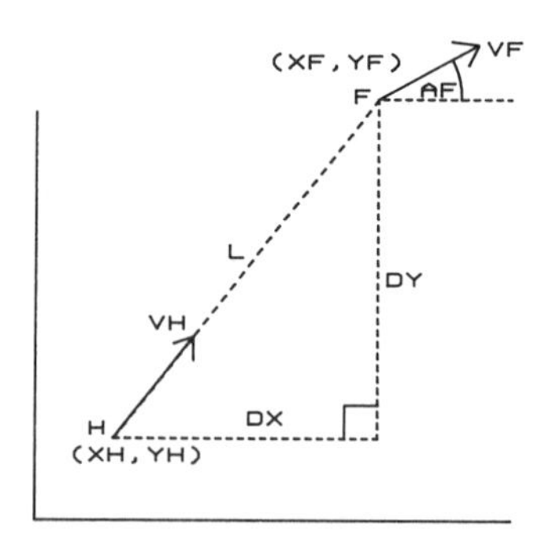

Fig. 3.23

We can now "update" the hound's position by using: XH=XH+M*DX and YH=YH+M*DY and draw the line between the hound's positions at the start and end of the time interval DT.

Similarly, since the fox is travelling along at a steady speed, we can update its position and draw the line that shows its progress in the same time interval.

```
  10 REM Prog.3.9 - Curve of pursuit
  49 :
  50 MODE 1 : PROCsetup
  60 OX = 0.1*SW : OY = 0.1*SH
  70 SX = 40 : SY = 40
  99 :
 100 XF = 1 : YF = 6
 110 VF = 1 : AF = PI/6
 120 XH = 0 : YH = 0 : VH = 1.5
 130 T = 0 : DT = 1 : HV = VF*COS(AF) : VV = VF*SIN(AF)
 140 REPEAT
 150  PROCchase
 160  T = T+DT
 170 UNTIL T>10
 490 END
 499 :
1300 DEF PROCchase
1310  DX = XF-XH : DY = YF-YH : L = SQR(DX*DX+DY*DY)
1320  X1 = XH : Y1 = YH : M = DT*VH/L : XH = XH+M*DX : YH = YH+M*DY
1330  X = XH : Y = YH :
1340  PROCjoin(X1,Y1, X,Y, 3)
1350  X1 = XF : Y1 = YF
1360  XF = XF+HV*DT : YF = YF+VV*DT
1370  X = XF : Y = YF
1380  PROCjoin(X1,Y1, X,Y, 3)
1390 ENDPROC
```

The output from this program is shown in Fig. 3.24. As previously now that

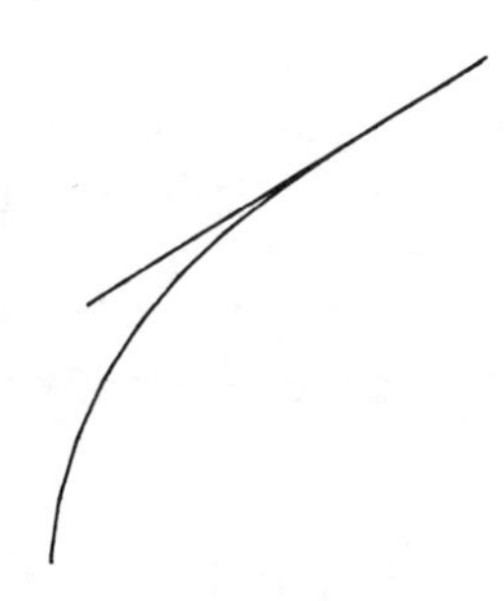

Fig. 3.24

the "theme" has been defined we can start to indulge in "variations".

How could you arrange for the program to stop when (and if) the hound catches the hare?

See if you can model the effect of one or other animal getting tired.

4

Curves — fancy

In this chapter we shall apply some of the techniques of the earlier chapters to a variety of problems concerning properties of curves. For example, we can start with exploring a few properties of the conics.

4.1 SOME PROPERTIES OF CONICS

4.1.1 The focus/directrix definition of a conic

In Chapter 2 we approached the circle and ellipse as the loci of internal points on a "slipping ladder". A more usual definition of a conic that covers both these curves, but also the parabola and hyperbola as well, is:

> Given a fixed point S and a line L, not passing through S, a conic is the locus of a point P such that its distance from S is always a constant multiple E of its distance from the line L.

If M is the foot of the perpendicular from P to the line L the SP=E.PM.

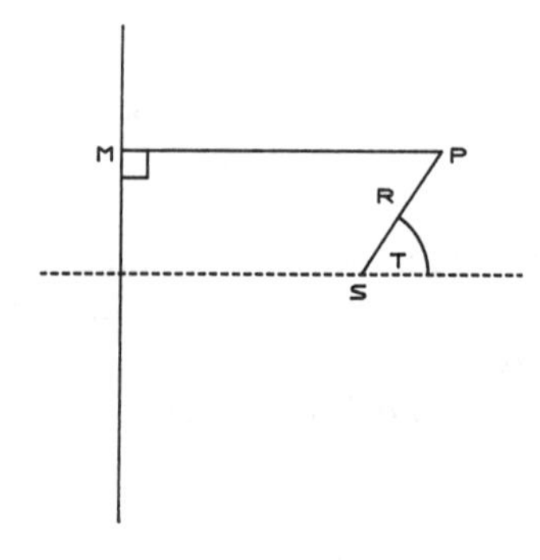

Fig. 4.1 — Focus/directrix definition of conic.

We can take the origin at S, make L be the line X=−A and use polar coordinates R and T to fix P. Then:

SP=R and PM=A+R*COS(T)

thus: SP=R=E.PM=E*A+E*R*COS(T)

i.e. R*(1−E*COS(T))=E*A

and R=E*A/(1−E*COS(T))

If we use T as a parameter then we can find the coordinates of P (relative to S as origin) from:

X=R*COS(T) Y=R*SIN(T)

In the following program (X1,Y1) are the coordinates of the previous position P′ of P. The origin is shifted to S and the line L and its perpendicular through S are drawn in red. For each value of T the lines SP and PM are drawn in yellow and a white line segment is drawn P′ to P to show the locus of P.

```
 10 REM Prog.4.1 - Focus/Directrix definition of conics
 49 :
 50 MODE 1 : PROCsetup
 60 OX = 0.5*SW : OY = 0.5*SH
 70 SX = 100 : SY = 100
 99 :
100 A = 2 : E = 0.5
110 PROCjoin(-5,0, 5,0, 1)
120 PROCjoin(-A,-5, -A,5, 1)
130 X1 = E*A/(1 - E) : Y1 = 0
140 FOR T = 0 TO 6.3 STEP PI/20
150  R = E*A/(1 - E*COS(T))
160  X = R*COS(T) : Y = R*SIN(T)
170  PROCjoin(0,0, X,Y, 2)
180  PROCjoin(X,Y, -A,Y, 2)
190  PROCjoin(X1,Y1, X,Y, 3)
200  X1 = X : Y1 = Y
210 NEXT T
490 END
499 :
500 DEF PROCsetup
510  SW = 1280 : SH = 1024
520  NC = 3 : FC = 3
590 ENDPROC
599 :
700 DEF PROCjoin(X1,Y1,X,Y,PC)
710  GCOL 0,PC
720  MOVE OX + X1*SX, OY + Y1*SY
730  DRAW OX + X*SX, OY + Y*SY
790 ENDPROC
799 :
```

Experiment with different values of E to produce ellipses, circles and hyperbolas. Can you explain why a pair of straight lines in the shape of an “X” appear when E is larger than 1? When E=1 the locus is supposed to be

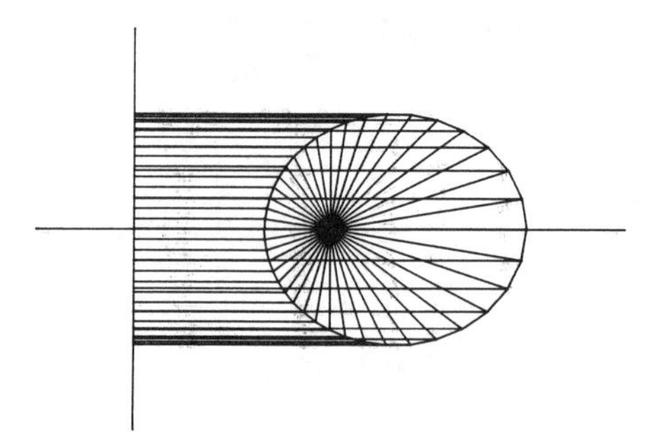

Fig. 4.2 — Output from program 4.1.

parabola, why will this program not draw it? Can you adapt the program to deal with this case?

4.1.2 Some string and two pins

One way of drawing an ellipse is to put two drawing-pins through a sheet of paper into a board and to tie each pin to the end of a piece of string. If you move a pen against the string so that it is taut then its path is supposed to be an ellipse.

Suppose the pins S and S′ are taken as the points (0,0) and (2*A,0) and that the string has length L (greater than 2*A). We can fix each position P of the pen by giving it polar coordinates (R,T) relative to S. If we use T as the parameter then we need to find an expression for R in terms of T.

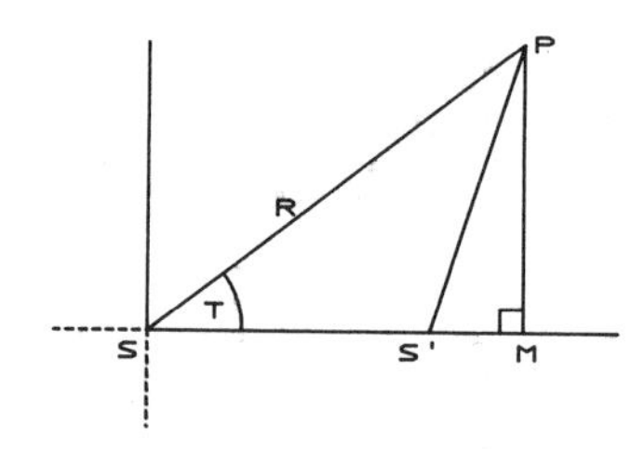

Fig. 4.3 — Diagram for string and 2 pins.

If M is the foot of the perpendicular from P to SS′ and the angle MPS′=U

then: PM=R*SIN(T)=(L−R)*COS(U)
and: SS′=2*A=SM+MS′=R*COS(T)+(L−R)*SIN(U)

We can eliminate U by squaring and adding to get:

(R*SIN(T))^2+(2*A−R*COS(T))^2=(L−R)^2

which simplifies to:

R=(4*A*A−L*L)/(4*A*COS(T)−2*L)

The cartesian coördinates (X,Y) of P are easily found from:

X=R*COS(T)   Y=R*SIN(T)

In the following program the axes of the ellipse are marked in red. The point P′ (X1,Y1) is the previous position of P. The lines SP and PS′ are marked in yellow and P′ is joined to P in white to show the locus.

```
 10 REM Prog.4.2 - String and 2 pins
 49 :
 50 MODE 1 : PROCsetup
 60 OX = 0.5*SW : OY = 0.5*SH
 70 SX = 100 : SY = 100
 99 :
100 A = 2 : L = 5
110 PROCjoin(-5,0, 5,0, 1)
120 PROCjoin(-A,-5, -A,5, 1)
130 OX = OX - A*SX
140 X1 = (4*A*A-L*L)/(4*A-2*L) : Y1 = 0
150 FOR T = 0 TO 6.3 STEP PI/20
160  R = (4*A*A-L*L)/(4*A*COS(T)-2*L)
170  X = R*COS(T) : Y = R*SIN(T)
180  PROCjoin(0,0, X,Y, 2)
190  PROCjoin(X,Y, 2*A,0, 2)
200  PROCjoin(X1,Y1, X,Y, 3)
210  X1 = X: Y1 = Y
220 NEXT T
490 END
```

What is the relationship between the length of string L and the eccentricity E

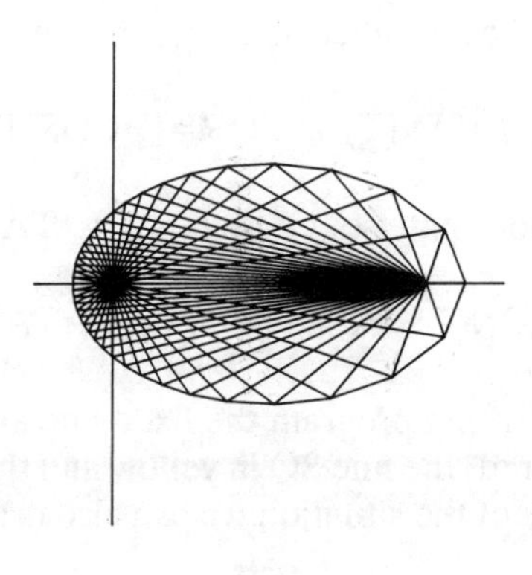

Fig. 4.4 — Output from program 4.2.

of an ellipse? What are the coordinates of P relative to the centre of the ellipse? Can you prove that the locus of P is an ellipse?

4.1.3 An envelope for a parabola

Lockwood gives a construction for enveloping a parabola in which S is a fixed point and there is a fixed line MN that does not pass through S. From S draw a line to any point Q on the line. From Q draw a line at right angles to SQ. These lines should form the envelope to a curve.

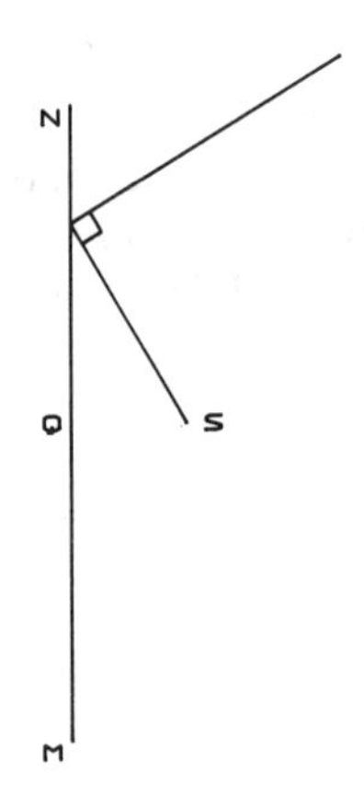

Fig. 4.5 — Set square construction.

We can first shift the origin to S(0,0) and make the line X=−A. Thus the point M(−A,0) lies on the line. Suppose SQ makes an angle T with the line Y=0 then:

MQ=A*TAN(T)

If P is distance L away from Q along the line perpendicular to SQ, and R is the foot of the perpendicular from P to the line MQ, then:

PR=L*SIN(T) and QR=L*COS(T)

Hence the coordinates of Q are (−A, A*TAN(T)) and those of P are:

(L*SIN(T)−A, L*COS(T)+A*TAN(T))

In the following program the fixed line and its perpendicular through S are drawn in red, the line SQ in yellow and the line QP in white. Because of the symmetry of the situation we can also draw in the reflections in the line Y=0.

```
10 REM Prog.4.3 - Envelope of parabola
49 :
```

```
 50 MODE 1 : PROCsetup
 60 OX = 0.5*SW : OY = 0.5*SH
 70 SX = 100 : SY = 100
 99 :
100 A = 1 : L = 10 : AS = PI/40
110 PROCjoin(-5,0, 5,0, 1)
120 PROCjoin(-A,-5, -A,512, 1)
130 FOR T = 0 TO ATN(5/A) STEP AS
140  AT = A*TAN(T) : LS = L*SIN(T) : LC = L*COS(T)
150  PROCjoin(0,0, -A,AT, 2)
160  PROCjoin(-A,AT, -A+LS,AT+LC, 3)
170  PROCjoin(0,0, -A,-AT, 2)
180  PROCjoin(-A,-AT, -A+LS,-AT-LC, 3)
190 NEXT T
490 END
```

The lines PQ are all tangents to a parabola — for each value of T how long

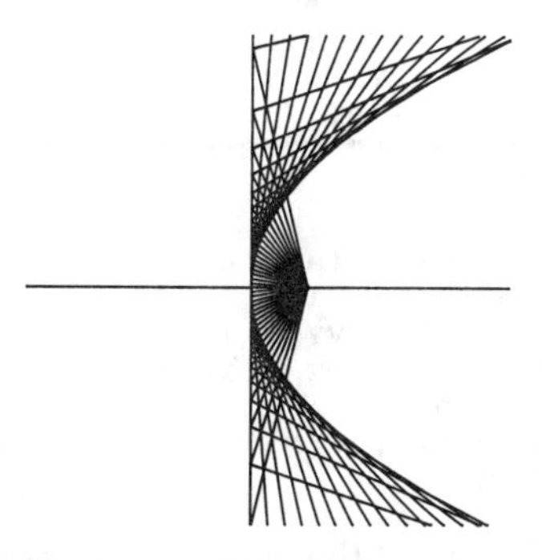

Fig. 4.6 — Output from program 4.3.

should L be if P is to just lie on the parabola?

4.1.4 Central orbits

Given the initial position and velocity of a body and some formula for calculating the force acting on the body at any position an important task is to determine the path the object describes. One such classic problem is given when the force at any position depends upon the distance of the object from a fixed point. A particularly important form of this is given by the so-called "inverse square law of attraction". In these circumstances we know that the possible orbits are all conics.

A simple, effective, but rather crude way of tackling this problem is to asume that the force remains constant over a small period of time DT.

In the following program the law of force is contained in line 140 and resolved into components of acceleration in line 150. Lines 160–190 "integrate" this twice on the assumption that the acceleration is constant over the small interval of time DT.

Line 200 draws a small section of arc and lines 210 and 220 "update" the position and the clock.

Initial values of the parameters are set in PROCinitialise. The program is terminated by pressing the ESCAPE key.

```
  10 REM Prog.4.4 - Central orbits
  49 :
  50 MODE 1 : PROCsetup
  60 OX = 0.5*SW : OY = 0.5*SH
  70 SX = 100 : SY = 100
  99 :
 100 PROCinitialise
 110 REPEAT
 120  R  = SQR(X1*X1 + Y1*Y1)
 130  C  = X1/R : S  = Y1/R
 140  F  = -MU/(R*R)
 150  AX = F*C : AY = F*S
 160  VX = VX + AX*DT
 170  VY = VY + AY*DT
 180  X  = X1 + VX*DT
 190  Y  = Y1 + VY*DT
 200  PROCjoin(X1,Y1, X,Y, 3)
 210  X1 = X : Y1 = Y
 220  T  = T  + DT
 300 UNTIL "Cows" = "Home"
 490 END
 499 :
1500 DEF PROCinitialise
1510  X1 = 4   : Y1 = 0
1520  VX = 0.5 : VY = 1
1530  T  = 0   : DT = 1/16
1540  MU = 10
1550  PROCdot(0,0, 3)
1590 ENDPROC
```

Investigate the shape of the curve for different values of the initial velocity

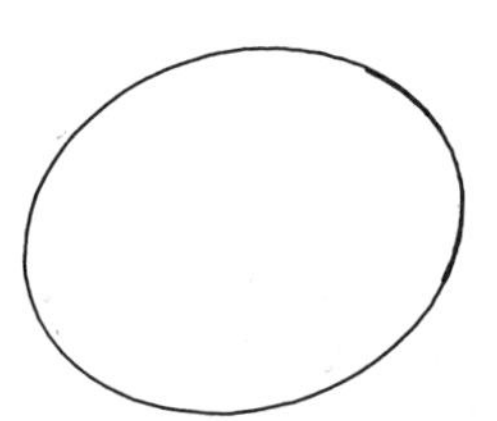

Fig. 4.7 — Output from program 4.4.

components VX,VY in line 1520. Can it be a circle?

This program works in simulated "real-time" — that is that for a fixed time interval the amount of arc of the orbit which is drawn depends upon the speed of the particle. From Kepler's observations if such a motion is elliptical then the object is supposed to "sweep out equal areas in equal times" — how could you check this? The theoretical orbit is an ellipse, so you could try to predict the orbit by choosing some suitable parameters and drawing an ellipse. Investigate other laws in line 140. You might like to try making improvements to the crude numerical algorithm used in lines 160–190.

4.1.5 An ellipse and its auxiliary circles

Another interesting source of ideas about geometric relations that we can explore with a microcomputer are books on technical (or engineering) drawing. For example, one conic property that is quoted is the relationship between an ellipse and two concentric circles. The smaller touches the ellipse at the ends of the minor axis and the larger touches at the ends of the major axis. These circles are called the **auxiliary** circles to the ellipse.

OP and OQ are radii of the two circles such that OPQ are collinear. A

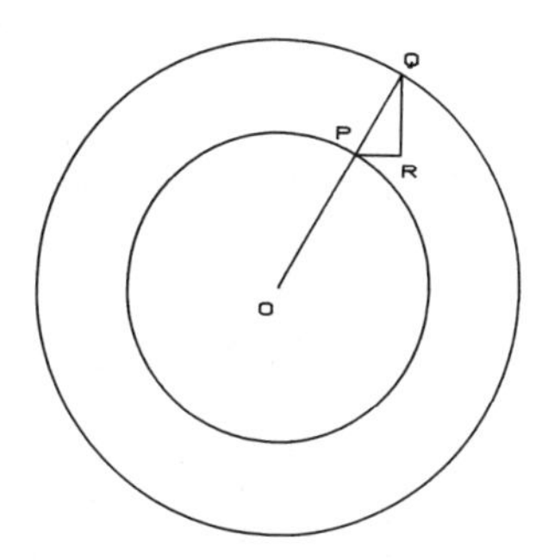

Fig. 4.8 — Ellipse and auxiliary circles.

right-angled triangle PQR is constructed on PQ as hypotenuse. The locus of R is the required ellipse.

In translating this into a program we have to use some form of coordinate representation and the result becomes obvious from our earlier work on circles and ellipses!

```
 10 REM Prog.4.5 - Auxiliary circles and ellipse
 49 :
 50 MODE 1 : PROCsetup
 60 OX = 0.5*SW : OY = 0.5*SH
 70 SX = 100 : SY = 100
 99 :
100 R1 = 3 : R2 = 5
110 PROCcircle(0,0,R1,1)
120 PROCcircle(0,0,R2,1)
130 FOR T = 0 TO 6.3 STEP PI/8
140  x1 = R1*COS(T) : y1 = R1*SIN(T)
150  x2 = R2*COS(T) : y2 = R2*SIN(T)
160  PROCjoin(0,0, x1,y1, 1)
170  PROCjoin(x1,y1, x2,y2, 2)
180  PROCjoin(x2,y2, x2,y1, 2)
190  PROCjoin(x2,y1, x1,y1, 2)
200 NEXT T
210 PROCellipse(0,0,R2,R1,3)
490 END
```

```
499 :
800 DEF PROCcircle(XC,YC,RC,PC)
810  PROCellipse(XC,YC,RC,RC,PC)
890 ENDPROC
899 :
900 DEF PROCellipse(XC,YC,RA,RB,PC)
910  X1 = RA : Y1 = 0
920  FOR P = 0 TO 6.3 STEP PI/20
930   X = RA*COS(P) : Y = RB*SIN(P)
940   PROCjoin(X1,Y1, X,Y, PC)
950   X1 = X : Y1 = Y
960  NEXT P
990 ENDPROC
```

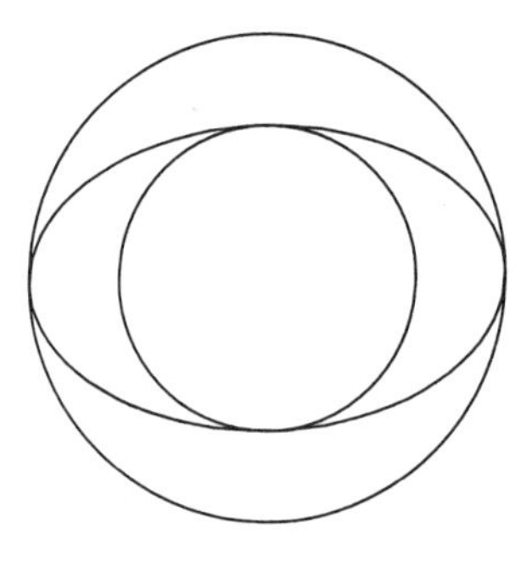

Fig. 4.9 — Output from auxiliary circles program 4.5.

4.2 TANGENTS, NORMALS, ARC-LENGTH, CURVATURE AND AREA

In this section we shall look at some examples of how the information used to draw curves in the last chapter can be applied to finding other characteristics.

4.2.1 Tangents to a circle

A general point on a circle radius R, centre the origin, is (Rcos(TH),Rsin(TH)) — how can we draw the tangent at this point? Obviously we can only draw some segment of the whole line which forms the tangent, so suppose we just want to draw a distance L in each direction.

Using similar triangles the coordinates of one end of the tangent are:

$$(R\cos(TH)+L\sin(TH),R\sin(TH)-L\cos(TH))$$

and those of the other end are:

$$(R\cos(TH)+L\sin(TH),R\sin(TH)+L\cos(TH))$$

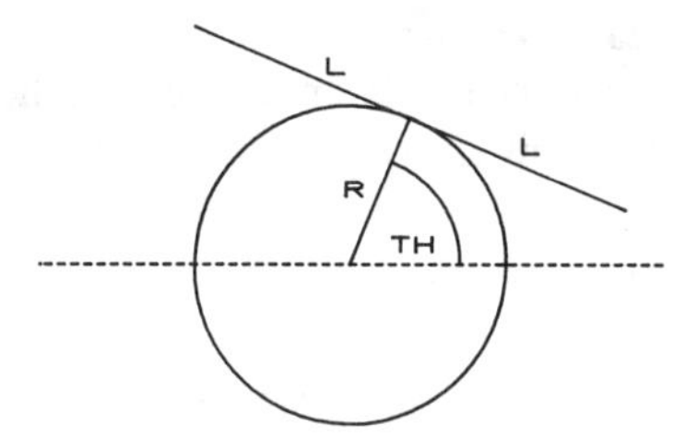

Fig. 4.10 — Tangents to a circle.

Thus a small program to draw such a set of tangents is:

```
 10 REM Prog.4.6 - tangents to a circle
 49 :
 50 MODE 1 : PROCsetup
 60 OX = 0.5*SW : OY = 0.5*SH
 70 SX = 100 : SY = 100
 99 :
100 R = 2 : L = 3
110 FOR T = 0 TO 6.3 STEP PI/10
120  C = COS(T) : S = SIN(T)
130  X1 = R*C + L*S : Y1 = R*S - L*C
140  X  = R*C - L*S : Y  = R*S + L*C
150  PROCjoin(X1,Y1, X,Y, FC)
160 NEXT T
490 END
```

How can we choose a value for L to ensure that each tangent is drawn just as far as some intersection with another tangent to give a star shape?

Suppose two such tangents make an angle 2*PH at the centre, then L=R*TAN(PH) — so we just need to ensure that PH is some multiple of the value PI/10 which is used for increments in TH:

```
100 M = 7 : PH = M*PI/10
105 R = 2 : L = R*TAN(PH)
```

What changes are needed to make these programs envelope an ellipse? If (A*COS(T), B*SIN(T)) is a point on an ellipse what is the gradient of the tangent at that point?

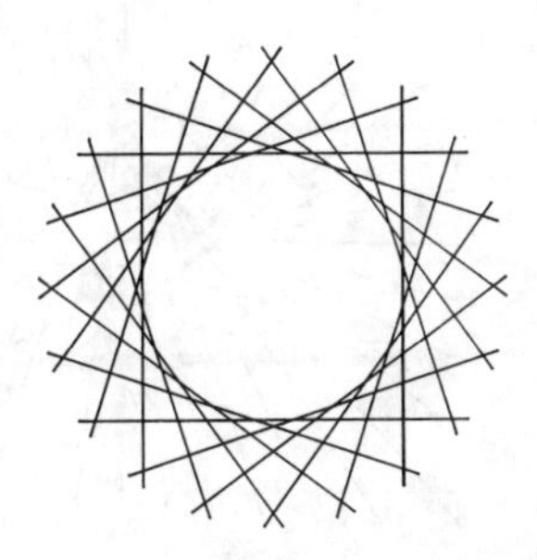

Fig. 4.11 — Output from program 4.6.

4.2.1 The normals to a curve

The normals to a curve will envelope another curve which is called its **evolute.** If a curve is given by the parametric forms:

x=FNx(t)   y=FNy(t)

then we can differentiate each with respect to the parameter t to obtain:

dx/dt=f=FNf(t)   dy/dt=g=FNg(t)

From these we can find the gradient of the tangent at (x,y) as g/f, and hence the gradient of the normal is −f/g. If we fix a constant length L to be marked off along the normal then we can find points distant L away from (x,y) along the normal and join them.

```
 10 REM Prog.4.7 - evolute of a curve
 49 :
 50 MODE 1 : PROCsetup
 60 OX = 0.5*SW : OY = 0.5*SH
 70 SX = 100 : SY = 100
 99 :
100 DEF FNx(t) =  A*COS(t)
110 DEF FNy(t) =  B*SIN(t)
120 DEF FNf(t) = -A*SIN(t)
130 DEF FNg(t) =  B*COS(t)
140 L = 5 : A = 3 : B = 2
150 t = 0 : x1 = FNx(t) : y1 = FNy(t)
160 PROCdot(x1,y1,2)
170 FOR t = 0 TO 6.3 STEP PI/20
180  x = FNx(t) : y = FNy(t)
190  PROCjoin(x1,y1, x,y, 2)
200  f = FNf(t) : g = FNg(t)
210  d = SQR(f*f + g*g)
220  x2 = x + g*L/d : y2 = y - f*L/d
230  x3 = x - g*L/d : y3 = y + f*L/d
240  PROCjoin(x2,y2, x3,y3, 3)
250  x1 = x : y1 = y
260 NEXT t
490 END
```

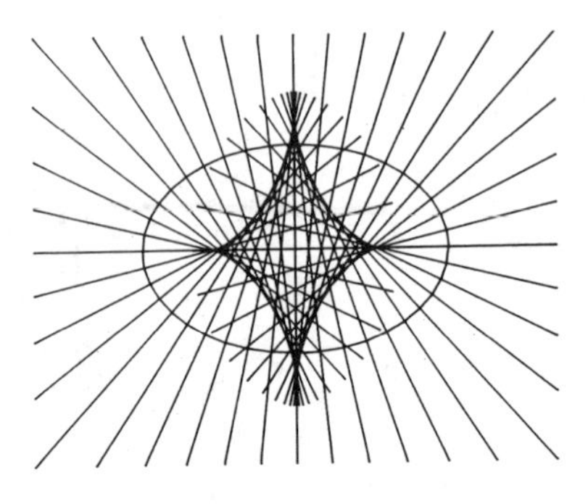

Fig. 4.12 — Output from program 4.7.

According to the theory the evolute is the locus of the centres of curvature of the base curve. Can you adapt the program to draw evolutes directly, i.e. without using envelopes?

4.2.3 Arc length and involutes

As the normal way of drawing curves on the computer screen is by approximating them with straight line segments it is easy to calculate the length of each segment and to accumulate their total as the curve is drawn. This gives an approximation to the length of the curved arc that will improve as the line segments are made shorter. In the following example we use the arc length to control the drawing of an associated curve. Start with some base curve and fix a point on it. Then, for any other point on the curve draw a tangent and mark off along it a length equal to the arc length on the curve between this point and the fixed point. The locus of such points is called the **involute** of the curve.

The variable S is used to accumulate the arc length, the functions FNf(t) and FNg(t) are the differentials of the functions FNx(t) and FNy(t) with respect to t. (x1,y1) are the coordinates of the last point reached on the base curve and (x,y) are those of the next point. (x2,y2) are the coordinates of the last point reached on the involute. The two curves and the tangents will all be drawn in different colours:

```
 10 REM Prog.4.8 - involute of a curve
 49 :
 50 MODE 1 : PROCsetup
 60 OX = 0.5*SW : OY = 0.7*SH
 70 SX = 100 : SY = 100
 99 :
100 DEF FNx(t) =  A*COS(t)
110 DEF FNy(t) =  B*SIN(t)
120 DEF FNf(t) = -A*SIN(t)
130 DEF FNg(t) =  B*COS(t)
140 A = 1 : B = 0.75 : s = 0
150 t = 0 : x1 = FNx(t) : y1 = FNy(t)
160 x2 = x1 : y2 = y1
170 PROCdot(x1,y1,2)
180 FOR t = 0 TO 6.3 STEP PI/20
190   x = FNx(t) : y = FNy(t)
200   f = FNf(t) : g = FNg(t)
210   d = SQR(f*f + g*g)
220   s = s + SQR((x-x1)^2 + (y-y1)^2)
230   x3 = x - f*s/d : y3 = y - g*s/d
240   PROCjoin(x1,y1, x3,y3, 1)
250   PROCjoin(x3,y3, x2,y2, 3)
260   PROCjoin(x1,y1, x,y, 2)
270   x1 = x :  y1 = y
280   x2 = x3 : y2 = y3
290 NEXT t
490 END
```

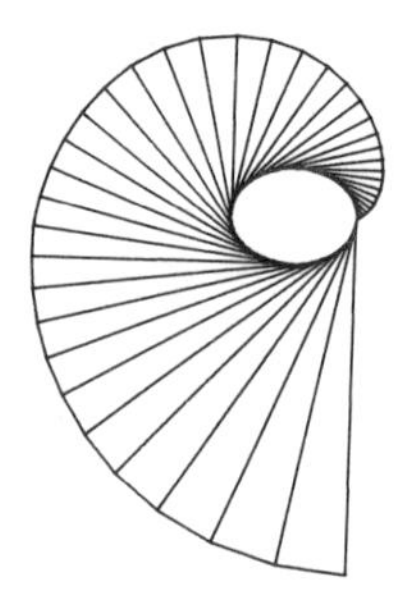

Fig. 4.13 — Output from program 4.8.

4.2.4 The circle of curvature

In the previous two examples we have used parametric representations of curves. Suppose we have a cartesian representation:

y=f(x)

and we denote the first and second derivatives of f(x) by f1(x) and f2(x). In this notation the curvature, k, of the curve at (x,f) is given by:

k=f2/SQR((1+f1 ↑ 2) ↑ 3)

and this is the reciprocal of the radius of curvature r. The centre (XC,YC) of the circle of curvature will lie on the normal to the curve at (x,f).

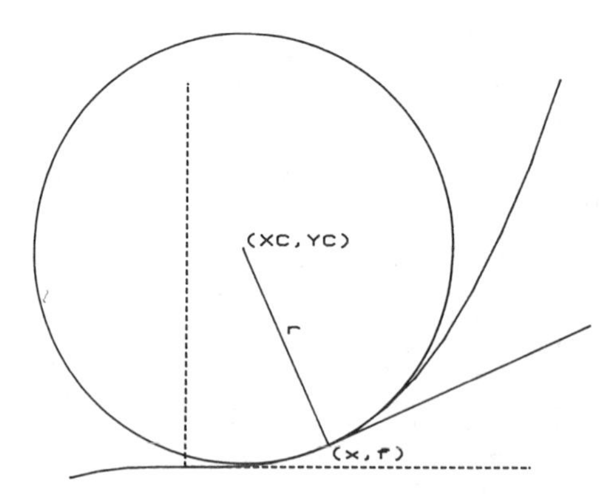

Fig. 4.14 — Circle of curvature.

The following program draws some red axes and the graph of the cartesian curve y=f(x) in yellow. The circle of curvature corresponding to a particular x value is then drawn in white using PROCcircle (which uses the "fast" circle algorithm):

```
 10 REM Prog.4.9 - circle of curvature
 49 :
 50 MODE 1 : PROCsetup
 60 OX = 0.1*SW : OY = 0.5*SH
 70 SX = 300 : SY = 300
 99 :
100 PROCjoin(0,0, 10,0, 1)
110 PROCjoin(0,-5, 0,5, 1)
120 DEF FNf(x)  =  COS(x) + SIN(x)
130 DEF FNf1(x) = -SIN(x) + COS(x)
140 DEF FNf2(x) = -COS(x) - SIN(x)
150 X1 = 0 : Y1 = 1
160 FOR X = 0 TO PI STEP 0.1
170  Y = FNf(X)
180  PROCjoin(X1,Y1, X,Y, 2)
190  X1 = X : Y1 = Y
200 NEXT X
210 x = PI/3
220 : REM Draw "analytic" circle
230 f1 = FNf1(x)
240 k = FNf2(x)/SQR((1+f1^2)^3)
250 r = 1/k : d = SQR(1+f1*f1)
260 XC = x-r*f1/d
270 YC = FNf(x)+r/d
280 PROCcircle(XC,YC,r,3)
490 END
499 :
800 DEF PROCcircle(XC,YC,RC,PC)
810  X3 = RC : Y3 = 0 : X1 = XC + X3 : Y1 = YC + Y3
820  S = SIN(PI/20) : C = COS(PI/20)
830  FOR P = 1 TO 40
840   X2 = X3*C - Y3*S : Y2 = X3*S + Y3*C
850   X = XC + X2 : Y = YC + Y2
860   PROCjoin(X1,Y1, X,Y, PC)
870   X3 = X2 : Y3 = Y2 : X1 = X : Y1 = Y
880  NEXT P
890 ENDPROC
```

Instead of using analytic expressions for the first and second derivatives

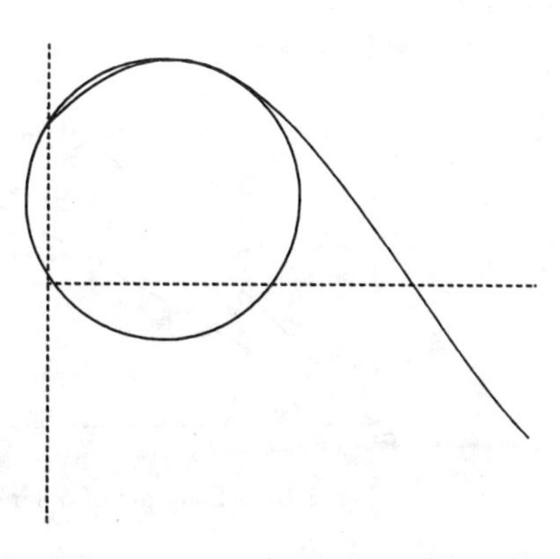

Fig. 4.15 — Output from program 4.9.

f1(x) and f2(x) we could investigate the use of numerical approximations.

If we define a small interval h then we can use finite approximations for f1 and f2 depending on h. A useful approximation for f1 is given by:

$$f1 = \frac{f(x+h) - f(x-h)}{2h}$$

and for f2 by:

$$f2 = \frac{f1(x+h) - f1(x-h)}{2h} = \frac{f(x+2h) - 2f(x) + f(x-2h)}{4h^2}$$

Using these approximations we can add the approximate circle of curvature in yellow by:

```
300 : REM Draw "numerical" circle
310 h = 1/8 : h2 = 2*h
320 f1 = (FNf(x+h)-FNf(x-h))/h2
330 f2 = (FNf(x+h2)-2*FNf(x)+FNf(x-h2))/(h2*h2)
340 k = f2/SQR((1+f1^2)^3)
350 r = 1/k : d = SQR(1+f1*f1)
360 XC = x-r*f1/d
370 YC = FNf(x)+r/d
380 PROCcircle(XC,YC,r,2)
```

The sensitivity of such an approximation can be tested by changing the value of h.

4.2.5 Arc length and area of a cardioid

We can use similar techniques to approximate the arc length and area contained within a closed curve in polar form. If (RO,TO) and (R,T) are the polar coordinates of successive points P,Q on a polar curve R=f(T) then we can approximate the element of arc-length ds of the curve by the length PQ and the area dA of the sector by the area of the triangle POQ.

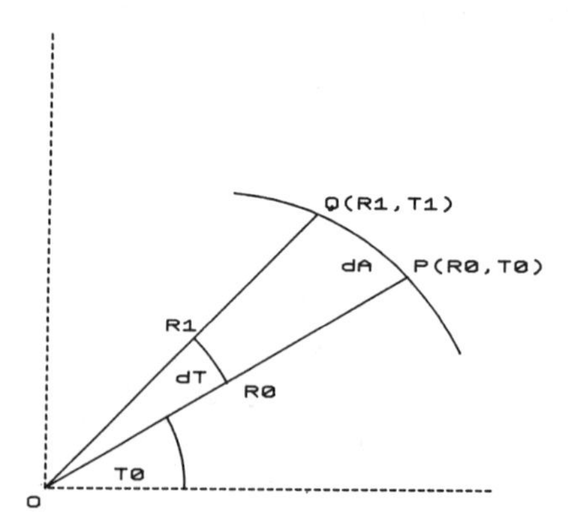

Fig. 4.16 — Diagram of polar curve.

```
10 REM Prog.4.10 - arc-length and area of a cardioid
49 :
50 MODE 1 : PROCsetup
```

```
 60 OX = 0.5*SW : OY = 0.5*SH
 70 SX = 100 : SY = 100
 99 :
100 DEF FNR(T) = 2*A*(1 - COS(T))
110 A = 1 : Arc = 0 : Area = 0
120 T1 = 0 : R1 = FNR(T1) : X1 = R1*COS(T1) : Y1 = R1*SIN(T1)
130 FOR T = 0 TO 6.3 STEP PI/20
140  R = FNR(T) : X = R*COS(T) : Y = R*SIN(T)
150  PROCjoin(X1,Y1, X,Y, FC)
160  Area = Area + 0.5*R*R1*SIN(T-T1)
170  Arc = Arc + SQR((X-X1)^2 + (Y-Y1)^2)
180  X1 = X : Y1 = Y : R1 = R : T1 = T
190 NEXT T
200 PRINT " Arc = "; Arc
210 PRINT " Area = "; Area
490 END
```

Compare these values with the analytic equivalents of 16*A for the arc length and 6*PI*A ↑ 2 for the area.

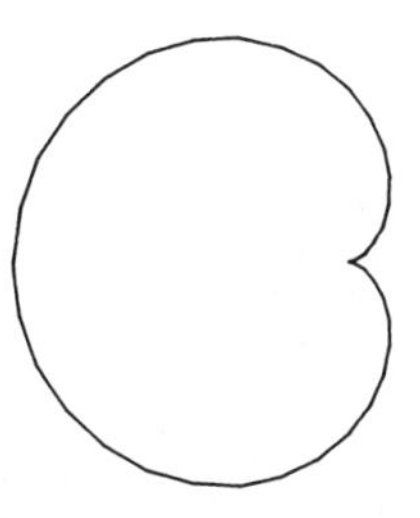

Fig. 4.17 — Output from program 4.10.

4.3 GOING ROUND IN CIRCLES

Now that we have met several examples of how to define procedures for drawing circles (as well as polygons and ellipses) we can explore a few of the many constructions that are based on circles.

4.3.1 Rotating arms

In this example we fix two circles. A point P starts at an angle P1 on the first circle (radius R1 and centre C1(XC1,YC1)) rotating with an angular velocity T1. A similar point Q starts at an angle P2 on the second circle (radius R2 and centre C2(XC2,YC2)) rotating with angular velocity T2. A point Z on the line PQ is defined by a factor F — if $0<F<1$ the point is inside the line-segment PQ. We require a program to explore the locus of Z for different values of the parameters P1,R1,XC1,YC1,T1,P2,R2, XC2,YC2,T2 and F. (There is no need for F to be fixed — it could vary with time!)

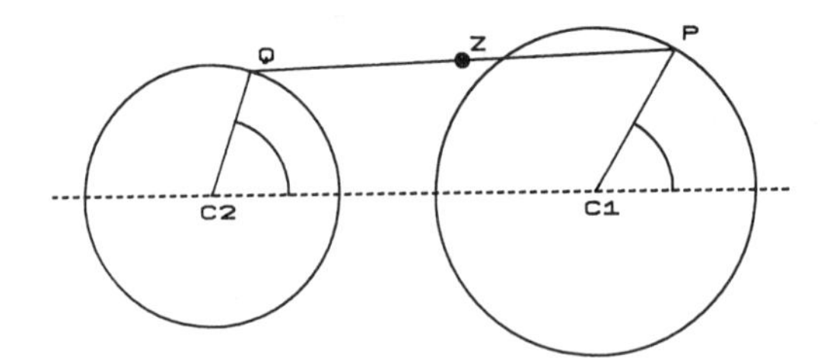

Fig. 4.18 — Arms diagram.

The following program draws the two base circles in yellow and then calculates the coordinates (x1,y1) of P and (x2,y2) of Q. The previous position of the point Z(X,Y) is kept stored as (X1,Y1). In this version the "linkage" C1.P.Q.C2 is drawn in red and the locus in white:

```
 10 REM Prog.4.11 - Rotating arms
 49 :
 50 MODE 1 : PROCsetup
 60 OX = 0.4*SW : OY = 0.5*SH
 70 SX = 100 : SY = 100
 99 :
100 R1 = 2.5 : XC1 = 4 : YC1 = 0 : T1 = 1 : P1 = 0
110 PROCcircle(XC1,YC1,R1,2)
120 R2 = 2 : XC2 = -2 : YC2 = 0 : T2 = 3 : P2 = PI/2
130 PROCcircle(XC2,YC2,R2,2)
140 F = 0.5
150 FOR T = 0 TO 6.3 STEP PI/20
160   A1 = T*T1 + P1 : A2 = T*T2 + P2
170   x1 = XC1 + R1*COS(A1) : y1 = YC1 + R1*SIN(A1)
180   x2 = XC2 + R2*COS(A2) : y2 = YC2 + R2*SIN(A2)
190   PROCjoin(XC1,YC1, x1,y1, 1)
200   PROCjoin(x1,y1, x2,y2, 1)
210   PROCjoin(x2,y2, XC2,YC2, 1)
220   X = (1-F)*x1 + F*x2 : Y = (1-F)*y1 + F*y2
230   IF T=0 THEN PROCdot(X,Y,3) ELSE PROCjoin(X1,Y1, X,Y, 3)
240   X1 = X : Y1 = Y
250 NEXT T
490 END
```

Explore the loci for different rates T1,T2 and different phase angles P1,P2. What happens if $F>1$ or $F<0$? Can you make F a function of T?

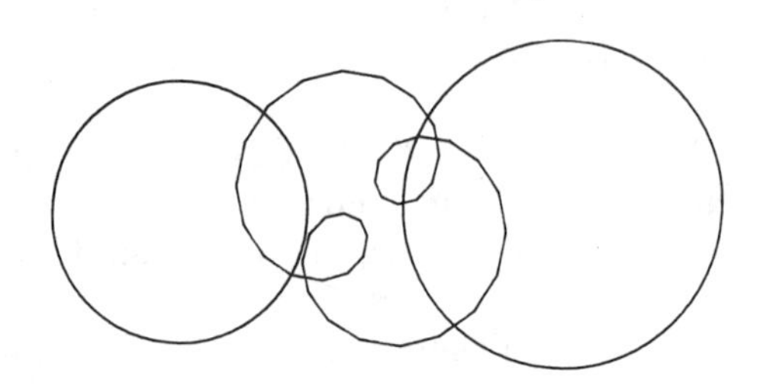

Fig. 4.19 — Output from "arms".

4.3.2 Circles through a point

Another Lockwood construction is as follows: fix a base circle, centre C (0,0) and radius R1 and some point S with coordinates (B,0). For each point P on the base circle a circle is drawn with P as centre to pass through S.

If R is the radius of the circle with centre at P — whose polar coordinates are R1 and T — then we can use the cosine rule on triangle PSC to give:

R*R−R1*R1+B*B−2*B*R1*COS(T)

and so we have R as a function of T.

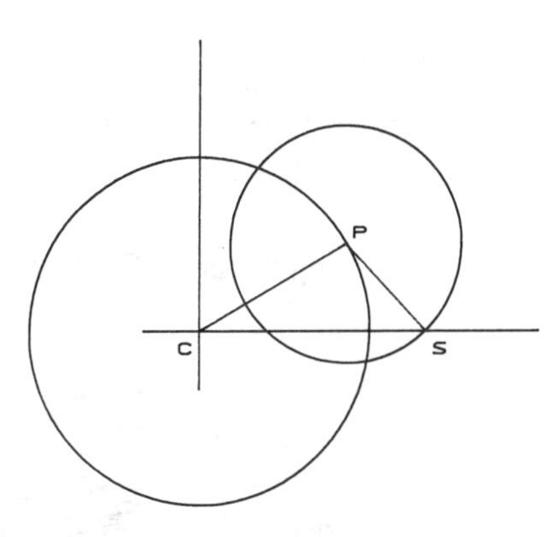

Fig. 4.20 — Diagram of circle & point.

In the following program many circles are to be drawn. In order to speed this up a procedure PROCtrig is introduced, which just stores a table of sines and cosines which can be 'looked-up' whenever a circle is to be drawn. The procedure PROCcircle(RR) draws a circle of radius RR with the current screen origin as centre and in the current graphics foregound colour. The base circle is drawn in yellow. For each value of the parameter T the value of the radius R is calculated and the origin shifted to the point P before the circle is drawn in white:

```
10 REM Prog.4.12 - Circle and point
49 :
50 MODE 1 : PROCsetup
60 OX = 0.5*SW : OY = 0.5*SH
```

```
  70 SX = 100 : SY = 100
  99 :
 100 R1 = 1.5 : B = R1 + 0.5
 110 PROCtrig : PROCcircle(0,0,R1,2)
 120 L = B*B + R1*R1 : M = 2*B*R1
 130 FOR T = 0 TO 6.3 STEP PI/20
 140  C = COS(T) : R = SQR(L - M*C)
 150  XC = R1*C : YC = R1*SIN(T)
 170  PROCcircle(XC,YC,R,3)
 180 NEXT T
 490 END
 499 :
 800 DEF PROCcircle(XC,YC,RC,PC)
 810  X1 = XC + RC : Y1 = YC
 820  FOR P = 1 TO 40
 830   X = XC + RC*C(P) : Y = YC + RC*S(P)
 840   PROCjoin(X1,Y1, X,Y, PC)
 850   X1 = X : Y1 = Y
 860  NEXT P
 890 ENDPROC
 899 :
1600 DEF PROCtrig
1610  DIM S(40),C(40)
1620  A = PI/20
1630  FOR P = 0 TO 40
1640   S(P) = SIN(P*A) : C(P) = COS(P*A)
1650  NEXT P
1690 ENDPROC
```

What is the effect of moving the point S inside the base circle? What shape is

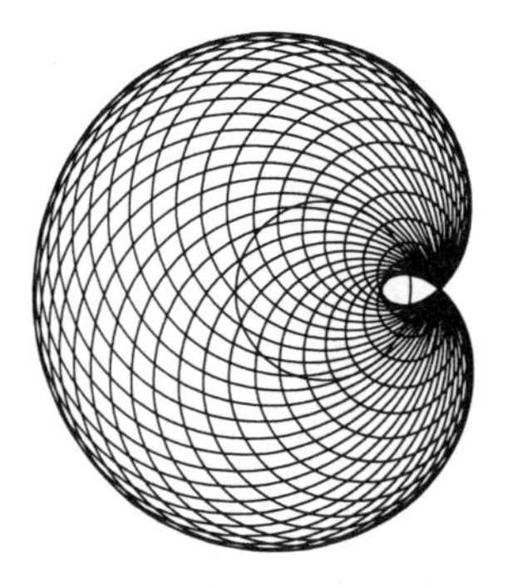

Fig. 4.21 — Output from circle/point.

formed when S is on the base circle? Could the base circle be generalised to a base ellipse?

4.3.3 Circles rolling on circles

A circle of radius R2 rolls on the outside of a base circle of radius R1 — what is the locus of a fixed point on the circumference? Suppose we take the centre of the base circle as origin O (0,0) and the circles start in contact at Q′ (R1,0). If we use T as parameter and P is the point of contact of the two circles then we must find the coordinates of the point Q on the rolling circle such that the arc length PQ of the rolling circle equals the arc length PQ′ of the base circle. If P is the size of the angle PCQ, where C is the centre of the rolling circle, then:

$$P*R2=T*R1 \text{ and so } P=T*R1/R2$$

The cartesian coordinates of C are ((R1+R2)*COS(T), (R1+R2)*SIN(T))

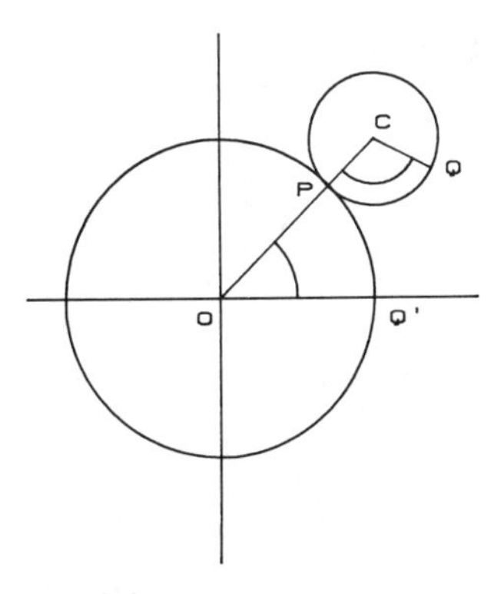

Fig. 4.22 — Diagram for epicycloids.

and the point Q has polar coordinates (R2,PI+T+P) referred to C as origin. Thus we can find the coordinates of Q referred to O as origin.

In the following program a **flag** is used to control what kind of display is produced. If the flag has the value TRUE then each position of the rolling circle will be shown in red, together with a radius from C to Q drawn in yellow. If the flag has the value FALSE then that part of the display will be skipped. Each point Q (X,Y) is joined to its previous value (X1,Y1) by a white line segment, and it is these which map out the locus of Q. Depending on the values of R1 and R2 it may be necessary to make T describe several cycles before the curve joints up again. Thus the main loop is an indefinite REPEAT-UNTIL loop which tests to see if we are very close to the original starting point Q′ whose coordinates are stored as (XO,YO):

```
 10 REM Prog.4.13 - Epicycloids
 49 :
 50 MODE 1 : PROCsetup
 60 OX = 0.5*SW : OY = 0.5*SH
 70 SX = 100 : SY = 100
 99 :
100 FLAG = FALSE
110 R1 = 2.4 : R2 = 1 : R = R1 + R2
120 PROCtrig : PROCcircle(0,0,R1,1)
130 X1 = R1 : Y1 = 0 : E = 0.01
140 XO = X1 : YO = Y1 : T = 0 : DT = PI/40
150 REPEAT
```

```
160  C = COS(T) : S = SIN(T)
170  XC = R*C : YC = R*S
190  IF FLAG THEN PROCcircle(XC,YC,R2,1)
200  P = R1*T/R2 : A = PI+T+P
210  X = XC + R2*COS(A) : Y = YC + R2*SIN(A)
220  IF FLAG THEN PROCjoin(XC,YC, X,Y, 2)
240  PROCjoin(X1,Y1, X,Y, 3)
250  X1 = X : Y1 = Y : T = T + DT
260 UNTIL (ABS(X1-XO)<E) AND (ABS(Y1-YO)<E) AND (T>DT)
490 END
```

The family of such curves are called **epicycloids** — explore the shape of such curves for different values of R1 and R2. What can you say about the way the ratio R1:R2 controls the curve?

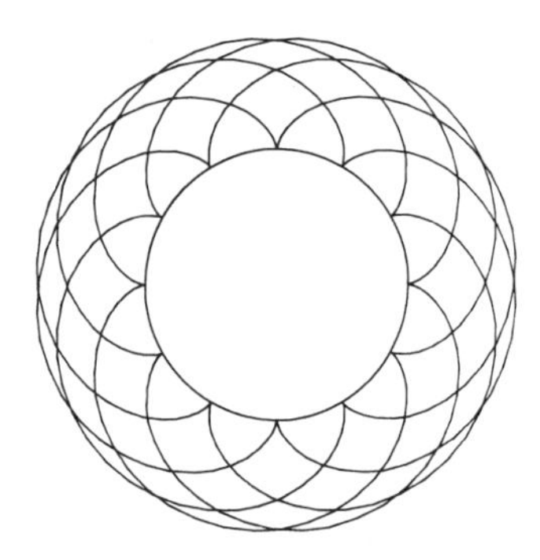

Fig. 4.23 — Output from epicycloids.

To make the circle rotate on the **inside** of the base circle we need to make only very slight changes.

```
110 R1 = 4 : R2 = 0.8 : R = R1-R2
200  P = R1*T/R2 : A = T-P
```

Can you make similar predictions about the shape of these curves — called **hypocycloids** — as the ratio of R1:R2 takes different values?
Both of these programs can be easily adapted to model the kind of designs

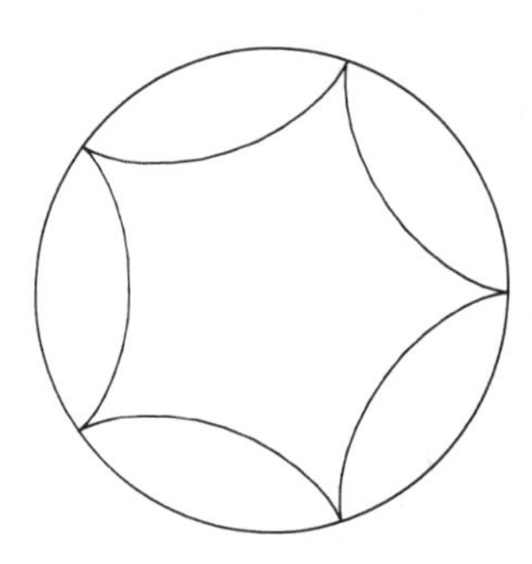

Fig. 4.24 — Output from hypocycloids.

obtained by using the commercially available **Spirograph** drawing kit. In this we take the locus of some point part way along a fixed radius of the rolling circle. If we make this point vary within a loop then a family of epi- or hypcycloids can be produced. The changes for the epicycloid are:

```
135 FOR F = 0.1 TO 1.01 STEP 0.1
136   R3 = F*R2 : X1 = R1+R2-R3 : Y1 = 0
210   X = XC + R3*COS(A) : Y = YC + R3*SIN(A)
265 NEXT F
```

and, additionally, for the hypocycloid:

```
136   R3 = F*R2 : X1 = R1-R2+R3 : Y1 = 0
```

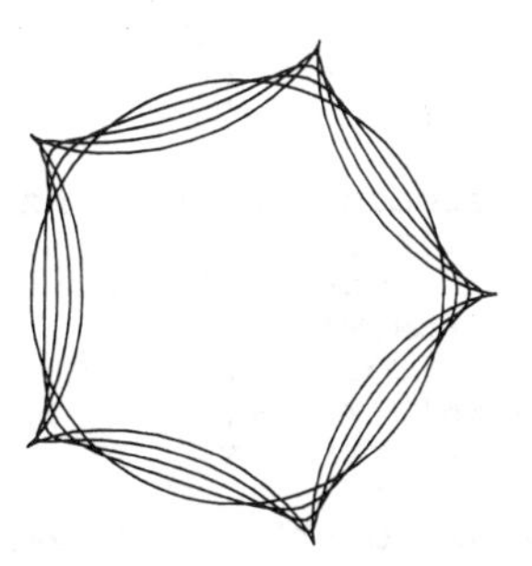

Fig. 4.25 — "Spirograph" output.

4.3.4 Inversion in a circle

If O is the centre of a circle of radius r then the points P and P′ are said to be inverse points with respect to the circle if:

$$OP.OP' = r^2$$

Some interesting results come from letting P describe some curve and considering the locus of its inverse point P′. The easiest kinds of curve for this purpose are ones which can express in **polar** form:

$$R = f(T)$$

where (R,T) are the polar coordinates of P using the centre of the circle O as origin.

The following program uses a defined function in line 100:

```
100 DEF FNR(T) = L/(1+E*COS(T))
```

which can be used to obtain ellipses, circles, parabolas and hyperbolas by suitable choices of L and E with the angle T varying between 0 and 2*PI.

The circle has radius RC and is drawn with the origin shifted to the centre of the screen. (X1,Y1) are the cartesian coordinates of the point on the defined curve with polar coordinates (R1,T), (X2,Y2) is the corresponding inverse point (R2,T) where R2=RC*RC/R1:

```
 10 REM Prog.4.14 - Inversion in a circle
 49 :
 50 MODE 1 : PROCsetup
 60 OX = 0.5*SW : OY = 0.5*SH
 70 SX = 100 : SY = 100
 99 :
100 DEF FNR(T) = L/(1+E*COS(T))
110 L = 2 : E = 1
120 RC = 2 : PROCcircle(0,0,RC,1)
130 R = RC*RC : REM Radius squared
140 T = 0 : R1 = FNR(T)
150 X1 = R1*COS(T) : Y1 = R1*SIN(T)
160 PROCdot(X1,Y1,2)
170 R2 = R/R1
180 X2 = R2*COS(T) : Y2 = R2*SIN(T)
190 PROCdot(X2,Y2,3)
200 FOR T = 0.05 TO 6.4 STEP PI/20
210  R1 = FNR(T)
220  X = R1*COS(T) : Y = R1*SIN(T)
230  PROCjoin(X1,Y1, X,Y, 2)
240  X1 = X : Y1 = Y
250  R2 = R/R1
260  X = R2*COS(T) : Y = R2*SIN(T)
270  PROCjoin(X2,Y2, X,Y, 3)
290  X2 = X : Y2 = Y
300 NEXT T
490 END
```

Try other kinds of function of T in line 100. Can you adapt the program to

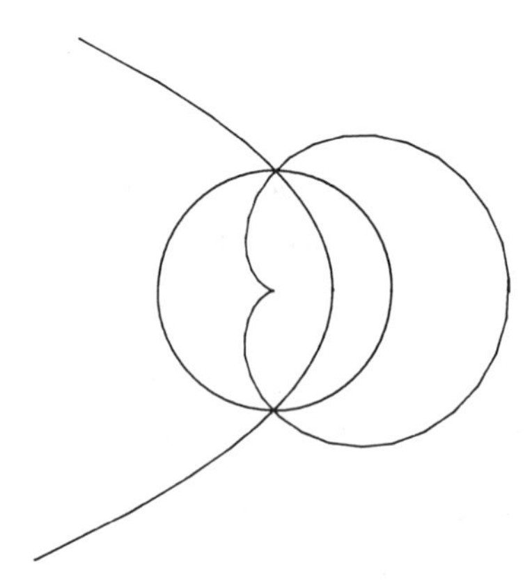

Fig. 4.26 — Output from Inversion.

handle other curves expressed in cartesian, rather than polar, form? What is the inverse of a straight line? How about performing inversion in curves other than circles? Try inversion in an ellipse.

In this technique a point on the circle corresponds to the **geometric** mean of the inverse points P and P′. With a couple of simple changes to this program we can change the model to a simple **reflection** in a circle in which case the point on the circle will correspond to the **arithmetic** mean of the reflected points P and P′ (i.e. R1+R2=2*RC). Another approach would be to use the **harmonic** mean (i.e. RC=2*R1*R2/(R1+R2).

4.4 TRIANGLES AND THEIR PROPERTIES

As an example of the use of the micro both as a "number cruncher" and as a "drawing machine" it is interesting to see how difficult (or easy) it might be to illustrate some of the major features of a triangle: medians, bisectors, altitudes, circumcircle etc. This is a good exercise both in finding mathematical forms appropriate for computer use and in writing a well-organised program. In this example we will use both the 2D coordinate geometry implicit in most micros' drawing commands and employ vector methods whenever they seem most appropriate.

The aim, then, is given the coordinates of the three vertices of a triangle to calculate and illustrate as many as possible of the triangle's principal features. To remind ourselves of that aim it helps to start by writing a "calling" program that names the procedures we need to define. Thus the following program will not run at all until the procedure "PROCtriangle" is defined:

```
 10 REM Prog.4.15 - Properties of a triangle
 49 :
 50 MODE 1 : PROCsetup
100 PROCtriangle(3,4, -4,0, 5,-2.5, 3)
110 PROCmedians(2)
120 PROCaltitudes(1)
130 PROCbisectors(3)
140 PROCcircumcircle(2)
150 PROCincircle(3)
160 PROC9point(1)
170 PROCexcircles(2)
490 END
499 :
```

The origin is in the middle of the screen and we "pass" the three pairs of coordinates as the "parameter list" to PROCtriangle. The first jobs, then, that this procedure needs to do are to store the coordinates in a usable form and to draw the triangle ABC:

```
1000 DEF PROCtriangle(x1,y1, x2,y2, x3,y3, PC)
1005  : REM Vertices; A,B,C
1010  Ax = x1 : Bx = x2 : Cx = x3
1020  Ay = y1 : By = y2 : Cy = y3
1030  PROCjoin(Ax,Ay, Bx,By, PC)
1040  PROCjoin(Bx,By, Cx,Cy, PC)
1050  PROCjoin(Cx,Cy, Ax,Ay, PC)
```

That was easy enough. Now we can use these coordinates to calculate the

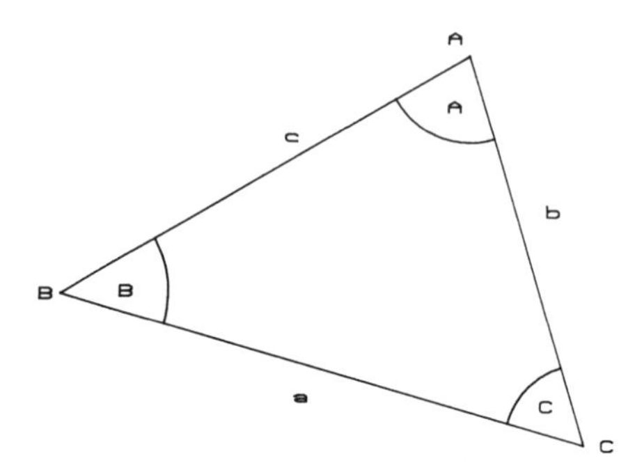

Fig. 4.27 — Diagram of triangle.

lengths a, b, c of the three sides using Pythagoras:

```
1055   : REM Sides; a,b,c
1060   a = SQR((Cx-Bx)^2 + (Cy-By)^2)
1070   b = SQR((Ax-Cx)^2 + (Ay-Cy)^2)
1080   c = SQR((Bx-Ax)^2 + (By-Ay)^2)
```

To find the angles we can use the cosine-rule and other trigonometric tricks:

```
1085   : REM Angles; A,B,C
1090   cosA = (b^2+c^2-a^2)/(2*b*c)
1100   sinA = SQR(1-cosA^2) : A = ACS(cosA)
1110   B = ACS((c^2+a^2-b^2)/(2*c*a))
1120   C = PI-A-B
```

Note here that "cosA" is the name of a variable. Keywords in BBC Basic are in capital letters and thus "ACS" is the Basic function that calculates the principal value of an angle (in radians) which has a given cosine.

Many remarkable expressions for statistics of the triangle are given in terms of the semi-perimeter, "s", and this is used next to calculate the area, "del", and the in-radius, "r". The circumradius, "R", is found from the sine-rule:

```
1125   : REM area, in- & circum-radii; del,r,R
1130   s = (a+b+c)/2
1140   del = SQR(s*(s-a)*(s-b)*(s-c))
1150   r = del/s : R = a/sinA/2
```

To find the coordinates of the foot D of an altitude of the triangle from A to BC it is convenient first to find the line equation of a side in the form:

$$1x+my+n=0$$

Then we can compute its intersection with the perpendicular from A:

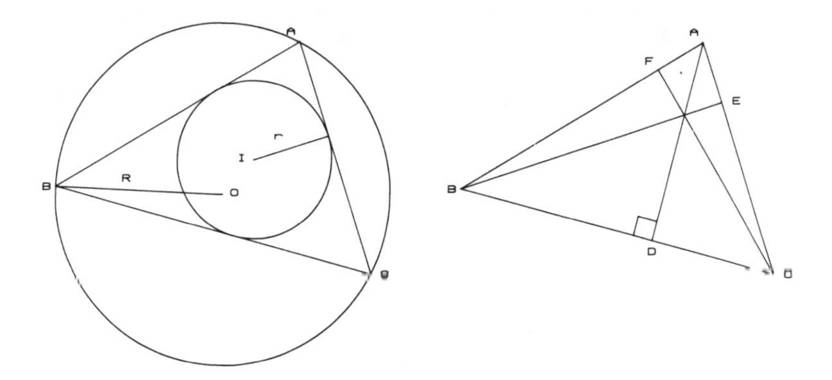

Fig. 4.28 — More details of triangle.

$$m(x-Ax)=1(y-Ay)$$

The following section does this job for all three altitudes:

```
1155   : REM feet of altitudes; D,E,F
1160   l1 = By-Cy : m1 = Cx-Bx
1170   n1 = Bx*Cy-Cx*By : p1 = l1^2+m1^2
1180   Dx = (m1^2*Ax-l1*m1*Ay-l1*n1)/p1
1190   Dy = (l1^2*Ay-l1*m1*Ax-m1*n1)/p1
1200   l2 = Cy-Ay : m2 = Ax-Cx
1210   n2 = Cx*Ay-Ax*Cy : p2 = l2^2+m2^2
1220   Ex = (m2^2*Bx-l2*m2*By-l2*n2)/p2
1230   Ey = (l2^2*By-l2*m2*Bx-m2*n2)/p2
1240   l3 = Ay-By : m3 = Bx-Ax
1250   n3 = Ax*By-Bx*Ay : p3 = l3^2+m3^2
1260   Fx = (m3^2*Cx-l3*m3*Cy-l3*n3)/p3
1270   Fy = (l3^2*Cy-l3*m3*Cx-m3*n3)/p3
```

After that relatively "heavy" piece of coordinate geometry we can return to the simpler task of finding the midpoints of the sides:

```
1275   : REM midpoints of sides; A_,B_,C_
1280   A_x = (Bx+Cx)/2 : A_y = (By+Cy)/2
1290   B_x = (Cx+Ax)/2 : B_y = (Cy+Ay)/2
1300   C_x = (Ax+Bx)/2 : C_y = (Ay+By)/2
```

Books, such as Coxeter's, usually refer to these points as A′, B′ and C′. Unfortunately the symbol, ′, cannot be used as part of a Basic variable and so I have used the "underline" symbol to give A__, B__ and C__.

To find the coordinates of the point P, where the bisector of angle A meets the side BC, we can use the result that P divides BC in the ratio AB:AC

```
1305   : REM angle bisectors; P,Q,R
1310   t1 = c/(b+c) : Px = (1-t1)*Bx + t1*Cx
1320   Py = (1-t1)*By + t1*Cy
1330   t2 = a/(c+a) : Qx = (1-t2)*Cx + t2*Ax
1340   Qy = (1-t2)*Cy + t2*Ay
```

```
1350  t3 = b/(a+b) : Rx = (1-t3)*Ax + t3*Bx
1360  Ry = (1-t3)*Ay + t3*By
```

The centroid G of the triangle is easily found from the averages of the

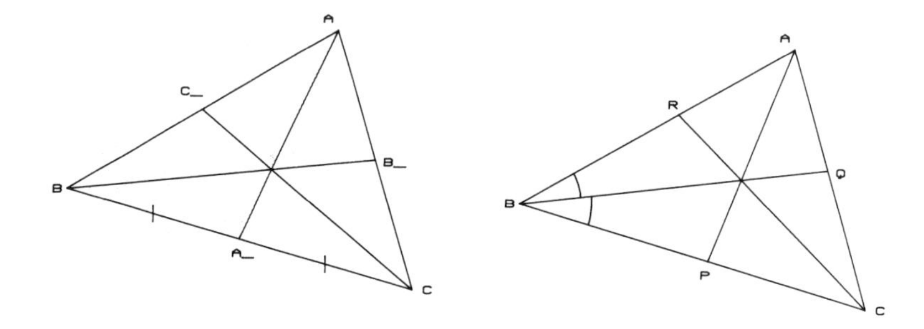

Fig. 4.29 — More diagrams of the triangle.

coordinates of A, B, C:

```
1365  : REM centroid; G
1370  Gx = (Ax+Bx+Cx)/3
1380  Gy = (Ay+By+Cy)/3
```

The coordinates of the circumcentre O are relatively heavy-going and can be found from the point of intersection of two of the perpendicular bisectors of the sides:

```
1385  : REM circumcentre; O
1390  den = Ay*(Cx-Bx) + By*(Ax-Cx) + Cy*(Bx-Ax)
1400  Ox = Ay*(Cx^2-Bx^2+Cy^2-By^2)
1410  Ox = Ox + By*(Ax^2-Cx^2+Ay^2-Cy^2)
1420  Ox = Ox + Cy*(Bx^2-Ax^2+By^2-Ay^2)
1430  Ox = Ox/den/2
1440  Oy = Ax*(Cy^2-By^2+Cx^2-Bx^2)
1450  Oy = Oy + Bx*(Ay^2-Cy^2+Ax^2-Cx^2)
1460  Oy = Oy + Cx*(By^2-Ay^2+Bx^2-Ax^2)
1470  Oy = -Oy/den/2
```

By contrast the incentre I has a much simpler form, and is just the weighted average of the three vertices:

```
1475  : REM incentre; I
1480  Ix = (a*Ax + b*Bx + c*Cx)/s/2
1490  Iy = (a*Ay + b*By + c*Cy)/s/2
```

To find the orthocentre H, the intersection of the altitudes, we could go through the process of calculating intersections again. However, there is a nice result, due to Euler, that connects the circumcentre O, the centroid G and the orthocentre H, which we can use as a short cut. This states that O, G,

H are collinear and that GH=2OG:

```
1495   : REM orthocentre; H
1500   Hx = 3*Gx-2*Ox : Hy = 3*Gy-2*Oy
```

Another interesting result, attributed to Feuerbach, is that the "orthic"

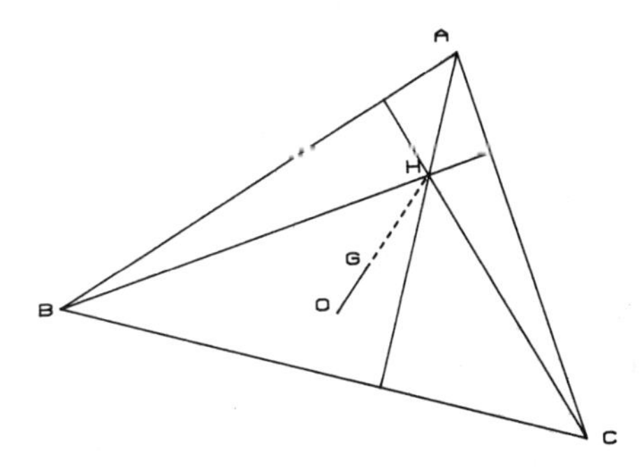

Fig. 4.30 — Further details about the triangle.

circle through the feet of the altitudes, D, E, F, also passes through some other interesting points. This so-called "nine-point" circle has a radius R9 which is half of the circumradius R and its centre N is the mid-point of the circumcentre O and the orthocentre H:

```
1505   : REM 9-point centre & radius; N,R9
1510   Nx = (Ox+Hx)/2 : Ny = (Oy+Hy)/2
1520   R9 = R/2
```

For the sake of completeness we can also find the centres and radii of the three "excircles" such as that which touches the side BC and the produced sides AB,AC:

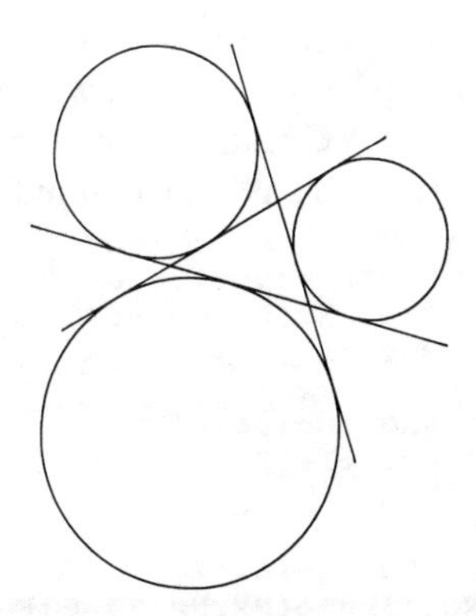

Fig. 4.31 — The 3 external circles to a triangle.

```
1525   : REM excentres & radii; Ia,Ib,Ic, ra,rb,rc
1530   Iax = ( a*Ax-b*Bx-c*Cx)/( a-b-c)
```

```
1540   Iay = ( a*Ay-b*By-c*Cy)/( a-b-c)
1550   Ibx = (-a*Ax+b*Bx-c*Cx)/(-a+b-c)
1560   Iby = (-a*Ay+b*By-c*Cy)/(-a+b-c)
1570   Icx = (-a*Ax-b*Bx+c*Cx)/(-a-b+c)
1580   Icy = (-a*Ay-b*By+c*Cy)/(-a-b+c)
1590   ra = del/(s-a) : rb = del/(s-b) : rc = del/(s-c)
1600 ENDPROC
1699 :
```

The welcome sight of the word "ENDPROC" means that we have concluded this set of triangle computations. In order to show how it may be used our main program also calls a number of other procedures which have yet to be defined. For example, line 110 draws the three medians in yellow using:

```
2000 DEF PROCmedians(PC)
2010   PROCjoin(Ax,Ay, A_x,A_y, PC)
2020   PROCjoin(Bx,By, B_x,B_y, PC)
2030   PROCjoin(Cx,Cy, C_x,C_y, PC)
2090 ENDPROC
2099 :
```

Similarly we can draw the altitudes:

```
2100 DEF PROCaltitudes(PC)
2110   PROCjoin(Ax,Ay, Dx,Dy, PC)
2120   PROCjoin(Bx,By, Ex,Ey, PC)
2130   PROCjoin(Cx,Cy, Fx,Fy, PC)
2190 ENDPROC
2199 :
```

and the internal angle bisectors:

```
2200 DEF PROCbisectors(PC)
2210   PROCjoin(Ax,Ay, Px,Py, PC)
2220   PROCjoin(Bx,By, Qx,Qy, PC)
2230   PROCjoin(Cx,Cy, Rx,Ry, PC)
2290 ENDPROC
2299 :
```

To draw the circles such as the circum-, in-, 9-point and ex-circles we need to include a familiar procedure to draw a circle specified by four parameters: the coordinates XC,YC of its centre, its radius RC and the "pen colour" PC. Once this circle routine has been defined the final touches are trivial indeed:

```
2300 DEF PROCcircumcircle(PC)
2310   PROCcircle(Ox,Oy,R,PC)
2390 ENDPROC
2399 :
2400 DEF PROCincircle(PC)
2410   PROCcircle(Ix,Iy,r,PC)
2490 ENDPROC
2499 :
2500 DEF PROC9point(PC)
2510   PROCcircle(Nx,Ny,R9,PC)
2590 ENDPROC
2599 :
2600 DEF PROCexcircles(PC)
2610   PROCcircle(Iax,Iay,ra,PC)
```

```
2620  PROCcircle(Ibx,Iby,rb,PC)
2630  PROCcircle(Icx,Icy,rc,PC)
2690 ENDPROC
```

All that we have done so far it to build up a tool-kit that can be used in the mensuration and illustration of triangles. The output from this particular test program is a mess because it tries to show too many things at once. More interesting results can be obtained, for example, by fixing B and C and exploring how one or more of the "key-points" (like O, G, H, I etc.), behaves as the position of A varies. As an exercise the reader is invited to trace out the loci of G and H as the point A moves round the circle centre O starting from a position in which ABC is equilateral.

As an example of an application we can try to demonstrate Wallace's theorem that the circumcircles of four triangles formed by any four straight lines all pass through a point:

As in our approach to the Pappus diagram in Section 2.13 we can define

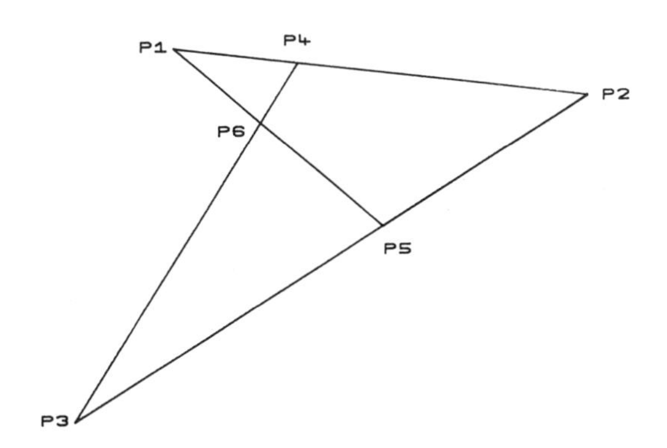

Fig. 4.32 — Diagram of Wallace's theorem.

just three fixed non-collinear points P1, P2 and P3 and define the other points to lie on the various line-segments by fixing parameters. Thus P4 lies on P1P2, and P5 lies on P3P4. P6 is the point of intersection of P1P5 with P3P4. Thus our program will need the PROCintersect procedure from Prog. 2.23 and we shall need some of the procedures just defined. In fact, we only need PROCcircumcircle, PROCcircle and just a small part of PROCtriangle:

```
 10 REM Prog.4.16 - Wallace's Theorem
 49 :
 50 MODE 1 : PROCsetup
 60 OX = 0.5*SW : OY = 0.5*SH
 70 SX = 100 : SY = 100
 99 :
100 P1x = -1.5 : P1y =  1.5
110 P2x =  2.5 : P2y =  1
120 P3x = -2.5 : P3y = -2
130 t = 0.3 : P4x = (1-t)*P1x + t*P2x : P4y = (1-t)*P1y + t*P2y
```

```
140 s = 0.6 : P5x = (1-s)*P3x + s*P2x : P5y = (1-s)*P3y + s*P2y
150 PROCintersect(P1x,P1y,P5x,P5y,P4x,P4y,P3x,P3y)
160 P6x = x : P6y = y
170 PROCjoin(P2x,P2y, P3x,P3y, 3)
180 PROCjoin(P3x,P3y, P4x,P4y, 3)
190 PROCtriangle(P1x,P1y, P2x,P2y, P5x,P5y, 3)
200 PROCcircumcircle(2)
210 PROCtriangle(P2x,P2y, P3x,P3y, P4x,P4y, 2)
220 PROCcircumcircle(2)
230 PROCtriangle(P3x,P3y, P5x,P5y, P6x,P6y, 2)
240 PROCcircumcircle(2)
250 PROCtriangle(P1x,P1y, P4x,P4y, P6x,P6y, 2)
260 PROCcircumcircle(2)
490 END
499 :
```

Once the definitions of PROCcircumcircle, PROCcircle and PROCintersect are provided considerable economies can be made in PROCtriangle, and the only additional procedure required is PROCcircumcircle:

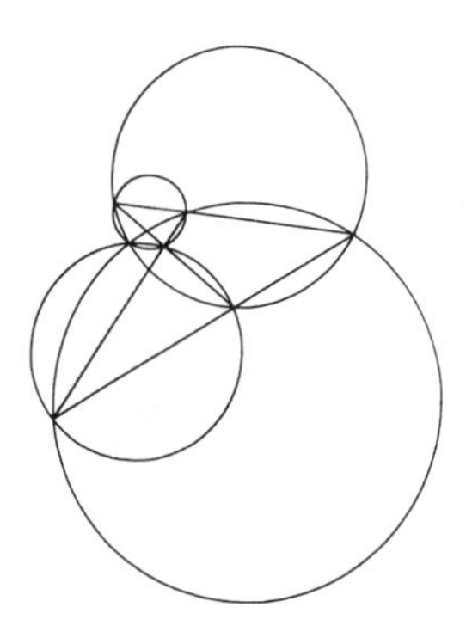

Fig. 4.33.

5

Transformations

Microcomputers are an extremely useful tool in exploring transformations geometry and hence many computer packages and programs of this type are already on the market. Similarly, there is quite a lot of literature already published on using micros in transformation geometry, for example in Chapter 17 of Oldknow and Smith's *Learning Mathematics with Micros* and in Chapter 4 of Oldknow's *Graphics with Microcomputers*. Thus this chapter does not aim to give comprehensive coverage of the topic but just to introduce one or two interesting ideas that may be explored. However, the opportunity is taken to extend one of the programs in *Oldknow and Smith* and to revise it in the structured form adopted in this text.

5.1 SIMPLE TRANSFORMATIONS REPEATED

The matrix for an anti-clockwise rotation through an angle A about the origin is:

$$M = \begin{bmatrix} \cos A & -\sin A \\ \sin A & \cos A \end{bmatrix}$$

Thus the point (X,Y) transforms to (XT,YT) where

XT=X*COS(A)−Y*SIN(A)

YT=X*SIN(A)+Y*COS(A)

The following program draws a 'flag' and some rotations:

```
 10 REM Prog.5.1 - Rotate a flag
 49 :
 50 MODE 1 : PROCsetup
 60 OX = 0.5*SW : OY = 0.5*SH
 70 SX = 100 : SY = 100
 99 :
100 DIM X(4),Y(4)
110 FOR I=1 TO 4
120    READ X(I),Y(I)
130 NEXT I
140 DATA 1,0, 5,0, 4,1, 3,0
```

```
 150 PROCdraw
 160 A = 30*(PI/180) : C = COS(A) : S = SIN(A)
 170 FOR I = 1 TO 11
 180   FOR J = 1 TO 4
 190     XT = X(J)*C - Y(J)*S
 200     YT = X(J)*S + Y(J)*C
 210     X(J) = XT : Y(J) = YT
 220   NEXT J
 230   PROCdraw
 240 NEXT I
 490 END
 499 :
5000 DEF PROCdraw
5010  X1 = X(1) : Y1 = Y(1)
5020  FOR P = 2 TO 4
5030   X = X(P) : Y = Y(P)
5040   PROCjoin(X1,Y1, X,Y, FC)
5050   X1 = X : Y1 = Y
5060  NEXT P
5090 ENDPROC
```

Adapt the program to enter in the coordinates of any figure and show them

Fig. 5.1 — Output of Prog. 5.1.

on the screen (some axes would help).

Lines 190, 210 perform the matrix multiplication for the transformation.

Adapt the program to perform other transformations, e.g. reflections, shears, stretches, etc.

Adapt the program so that you can enter your own matrix and examine the effect.

Extend the program to consider the product of two different transformations. Can you compute areas?

More general transformations can be studied by moving the origin between transformations — try to perform a rotation about, say, (3,3) or a reflection in the line $x + y = 5$

You might try to extend the program to explore inverses, e.g. (a) show a target shape, (b) transform it and show its image and (c) invite the user to specify a transformation, or its matrix, that will map the image back onto the target.

5.2 ROTATING RECTANGLES

Take a square, centre the origin, with one corner at (X,Y) — where are the other four corners? We could use the matrix for a rotation of 90 degrees, or just similar traingles to find them as:

$$(-Y,X), (-X,-Y) \text{ and } (Y,-X)$$

If the point (X,Y) is specified as being at a diagonal distance L from the

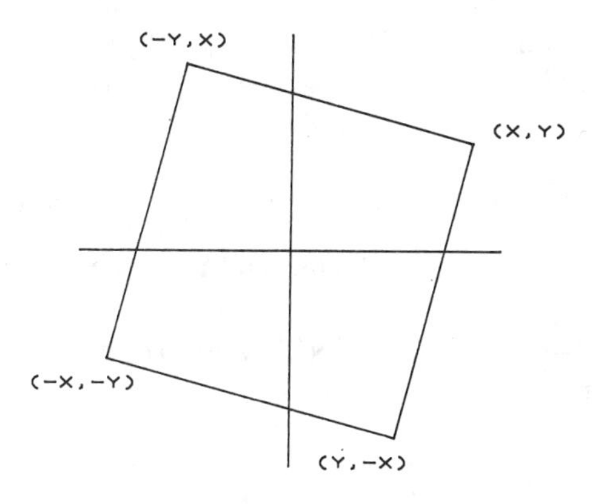

Fig. 5.2 — Diagram of rotated square.

origin, where the diagonal makes an angle T with the x-axis then X=L*COS(T) and Y=L*SIN(T) — thus we can make the square 'spin' by varying T in a loop:

```
 10 REM Prog.5.2 - rotating squares
 49 :
 50 MODE 1 : PROCsetup
 60 OX = 0.5*SW : OY = 0.5*SH
 70 SX = 100 : SY = 100
 99 :
100 L = 5
110 FOR T = 0 TO PI/2 STEP PI/20
120  X = L*COS(T) : Y = L*SIN(T)
130  PROCjoin(X,Y, -Y,X, FC)
140  PROCjoin(-Y,X, -X,-Y, FC)
150  PROCjoin(-X,-Y, Y,-X, FC)
160  PROCjoin(Y,-X, X,Y, FC)
170 NEXT T
490 END
```

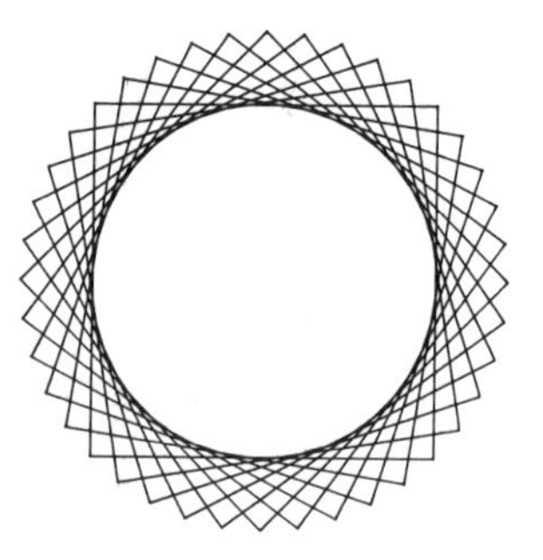

Fig. 5.3 — Output from Prog. 5.2.

Suppose we try to extend the idea to a rectangle whose diagonals cross at an angle A (which is 90 degrees in the case of a square). If (X,Y) is one corner then certainly (−X,−Y) is the diagonally opposite corner, but where are the other two? Again we can use trignometry or the rotation matrix for an angle A to find the coordinates:

$$\begin{bmatrix} COS(A) & -SIN(A) \\ SIN(A) & COS(A) \end{bmatrix} \begin{bmatrix} X \\ Y \end{bmatrix} = \begin{bmatrix} X*C-Y*S \\ X*S+Y*C \end{bmatrix}$$

where C=COS(A), S=SIN(A) and so only small changes are needed to the program:

```
 10 REM Prog.5.3 - Rotating rectangles
 49 :
 50 MODE 1 : PROCsetup
 60 OX = 0.5*SW : OY = 0.5*SH
 70 SX = 100 : SY = 100
 99 :
100 L = 50
115 A = PI/3 : S = SIN(A) : C = COS(A)
120 FOR T = 0 TO PI STEP PI/20
130  X = L*COS(T) : Y = L*SIN(T)
140  PROCjoin(X,Y, X*C-Y*S, X*S+Y*C, FC)
150  PROCjoin(X*C-Y*S, X*S+Y*C, -X,-Y, FC)
160  PROCjoin(-X,-Y, -X*C+Y*S,-X*S-Y*C, FC)
170  PROCjoin(-X*C+Y*S,-X*S-Y*C, X,Y, FC)
180 NEXT T
490 END
```

Did you expect the rectangles to envelope an ellipse in the way the squares evelope a circle? What can you do to either program to envelope an ellipse?

5.3 TESSELLATION OF THE PLANE

In his article "In Praise of Geometry" (Chapter 1 of the Schools Councils', *The Mathematics Curriculum — Geometry*), W. W. Sawyer takes an example of patterns produced from a simple shape like a capital letter "M" with the right-hand "leg" removed. This shape is first rotated through right-angles and then repeatedly reflected. We can build up this approach with the use of the micro:

```
 10 REM Prog 5.4 - Sawyer pattern 1
 50 MODE 1 : PROCsetup
 60 OX = 0.5*SW : OY = 0.5*SH
 70 SX = 80 : SY = 80
 99 :
100 L = 1 : L2 = 2*L
110 PROCjoin(0,0, 0,L2, FC)
120 PROCjoin(0,L2, L,L, FC)
130 PROCjoin(L,L, L2,L2, FC)
490 END
```

This first program just draws the basic shape made from three lines.

Fig. 5.4 — Output from Prog. 5.4.

In order to rotate the shape we need to store the coordinates and perform successive rotations:

```
  10 REM Prog.5.5 - Sawyer pattern 2
  49 :
  50 MODE 1 : PROCsetup
  60 OX = 0.5*SW : OY = 0.5*SH
  70 SX = 80 : SY = 80
  99 :
 100 L = 1 : L2 = 2*L
 110 N = 4
 120 DIM X(N),Y(N)
 130 X(1) = 0  : Y(1) = 0
 140 X(2) = 0  : Y(2) = L2
 150 X(3) = L  : Y(3) = L
 160 X(4) = L2 : Y(4) = L2
 170 FOR G = 1 TO 4
 180  PROCdraw
 190  PROCrotate
 200 NEXT G
 490 END
 499 :
5100 DEF PROCrotate
5110  FOR P = 1 TO N
5120   X = -Y(P)
5130   Y(P) = X(P)
5140   X(P) = X
5150  NEXT P
5190 ENDPROC
```

In this case PROCrotate just performs the special case of a 90-degree rotation.

We can show the bounding square by adding:

```
210 S = 2*L2
220 PROCjoin(S,0, 0,S, 1)
230 PROCjoin(0,S, -S,0, 1)
240 PROCjoin(-S,0, 0,-S, 1)
250 PROCjoin(0,-S, S,0, 1)
```

We can now tidy up the program in readiness for the reflective patterns:

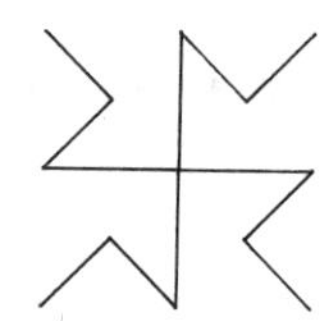

Fig. 5.5 — Output from Prog. 5.5.

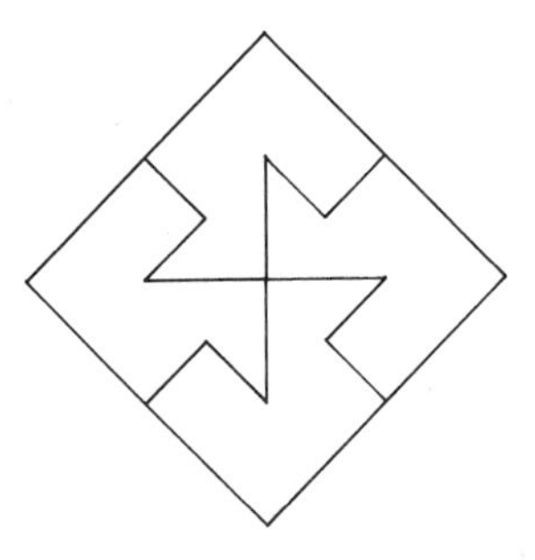

Fig. 5.6 — As Fig. 5.5 but with bounding sqaure.

```
  10 REM Prog.5.6 - Sawyer pattern 3
  49 :
  50 MODE 1 : PROCsetup
  60 OX = 0.5*SW : OY = 0.5*SH
  70 SX = 80 : SY = 80
  99 :
 100 PROCdefine
 110 PROCpattern
 490 END
 499 :
1000 DEF PROCdefine
1010  L = 1
1020  N = 4 : DIM X(N),Y(N)
1030  FOR P = 1 TO N
1040   READ X,Y
1050   X(P) = X*L : Y(P) = Y*L
1060  NEXT P
```

```
1070  DATA 0,0, 0,2, 1,1, 2,2
1090 ENDPROC
1099 :
1200 DEF PROCpattern
1210  FOR M = 1 TO 4
1220   PROCdraw
1230   PROCrotate
1240  NEXT M
1290 ENDPROC
```

Assuming, of course, that PROCdraw and PROCrotate are defined as previously.

Now we can implement the reflections by adding a procedure to perform a reflection in the line y = −x which we can use with shifts of origin to generate the repeated pattern.

```
  10 REM Prog.5.7 - Sawyer pattern 4
  49 :
  50 MODE 1 : PROCsetup
  60 XL = 0.4*SW : YL = 0.4*SH
  70 SX = 16 : SY = 16
  99 :
 100 PROCdefine
 110 FOR F = -2 TO 2
 120  OX = XL + 2*L*SX : OY = YL + (2+F*8)*L*SX
 130  FOR G = 1 TO 5
 140   PROCpattern
 150   PROCreflect
 160   OX = OX + 4*L*SX : OY = OY + 4*L*SY
 170  NEXT G
 180 NEXT F
 490 END
 499 :
1400 DEF PROCreflect
1410  FOR P = 1 TO N
1420   X = -Y(P)
1430   Y(P) = -X(P)
1440   X(P) = X
1450  NEXT P
1490 ENDPROC
```

Line 160 performs the translation by a shift of origin.

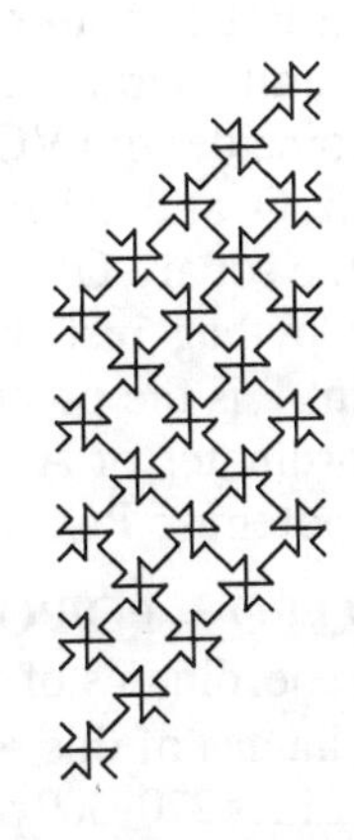

Fig. 5.7 — Output from Prog. 5.7.

See if you can adapt the idea to make tessellations from different basic shapes. According to the theory there are just 17 classes of such "wallpaper" designs — try varying the transformations used in making the pattern to model some of the other 16 classes.

5.4 PYTHAGORAS BY TRANSFORMATIONS

One of the few well documented uses of transformation geometry is in one method of proving Pythagoras' theorem. Figure 5.8 shows a diagram of a right-angled triangle POQ with squares drawn on each side in which the square on the hypotenuse is divided into two rectangles by the perpendicular from Q to PQ, produced to meet the futher side of the square at B.

The proof is based upon mapping each of the two rectangles onto each of the squares on the sides by a sequence of area-preserving transformations

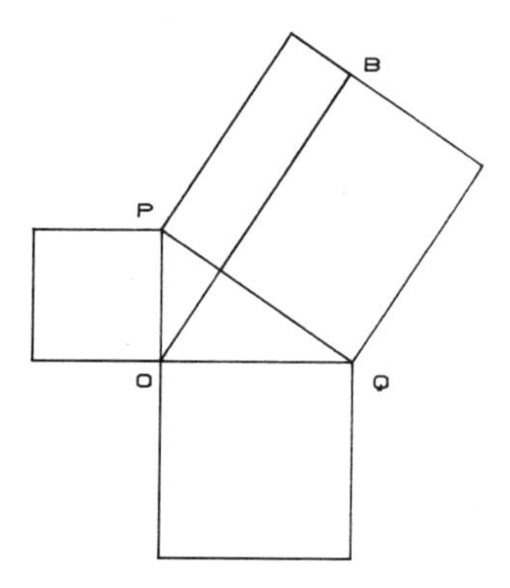

Fig. 5.8 — Pythagoras' diagram.

(i.e. the isometries — rotations, reflections and translations — together with shears). If we label the upper rectangle as P1, P2, P3, P4 with P1 initially at P we can first shear along P1P2 to make P4 coincide with O. Then we can rotate the resulting parallelogram about O anti-clockwise through 90 degrees to make P1P4 coincide with VO. Finally, we can shear along P1P4 to make P3 coincide with O and P2 coincide with W, thus mapping the rectangle APRB onto the square OVWP by two shears and a rotation.

In Prog. 5.8 PROCpythag just draws the outline of the Pythagoras diagram. O is the origin, P is the point (0,H) and Q is the point (W,0). In order to locate the coordinates of A and B we need to know the ratio in which A divides the hypotenuse PQ. Suppose PA = t.PQ then:

$$PA/PO = t.PQ/PO = PO/PQ \text{ and so } t = (PO)\uparrow 2/(PQ)\uparrow 2$$

and so we can find the coordinates of the the points P, R, B and A. Lines 170–220 store the coordinates of this rectangle and draw it. Lines 230, 260 perform the first shear. Lines 270–300 perform the rotation about O. Finally Line 310 does the concluding shear:

```
  10 REM Prog.5.8 - Illustrate proof of Pythagoras
  20 REM by 2 shears and a rotation
  49 :
  50 MODE 1 : PROCsetup
  70 SX = 100 : SY = 100
  99 :
 100 : REM Draw "pythagoras"
 110 W = 3 : H = 2
 120 OX = (H+4)*SX : OY = (W+1)*SY
 130 PROCpythag
 140 : REM store initial area
 150 t = H^2/(W^2+H^2)
 160 PROCjoin(0,0, H+t*W,W+(1-t)*H, 2)
 170 DIM X(4),Y(4)
 180 X(1) = 0 : Y(1) = H
 190 X(2) = H : Y(2) = W+H
 200 X(3) = H+t*W : Y(3) = W+(1-t)*H
 210 X(4) = t*W : Y(4) = (1-t)*H
 220 PROCdraw
 225 : REM shear along P1P2 to make P4 coincide with O
 230 X(3) = X(3)-X(4) : Y(3) = Y(3)-Y(4)
 240 X(4) = 0 : Y(4) = 0
 250 PRINT "Press any key" : INPUT K$
 260 PROCdraw
 265 : REM rotate 90 deg anticlockwise about O
 270 FOR I = 1 TO 4
 280  T = X(I) : X(I) = -Y(I) : Y(I) = T
 290 NEXT I
 295 PRINT "Press any key" : INPUT K$
 300 PROCdraw
 305 : REM shear along x-axis to make sides vertical
 310 X(2) = X(2)-X(3) : X(3) = 0
 315 PRINT "Press any key" : INPUT K$
 320 PROCdraw
 390 END
 499 :
4000 DEF PROCpythag
4010  PROCjoin(0,0, W,0, FC)
4020  PROCjoin(W,0, 0,H, FC)
4030  PROCjoin(0,H, 0,0, FC)
4040  PROCjoin(0,0, -H,0, 1)
4050  PROCjoin(-H,0, -H,H, 1)
4060  PROCjoin(-H,H, 0,H, 1)
4070  PROCjoin(0,H, H,W+H, 1)
4080  PROCjoin(H,W+H, W+H,W, 1)
4090  PROCjoin(W+H,W, W,0, 1)
4100  PROCjoin(W,0, W,-W, 1)
4110  PROCjoin(W,-W, 0,-W, 1)
4120  PROCjoin(0,-W, 0,0, 1)
4190 ENDPROC
```

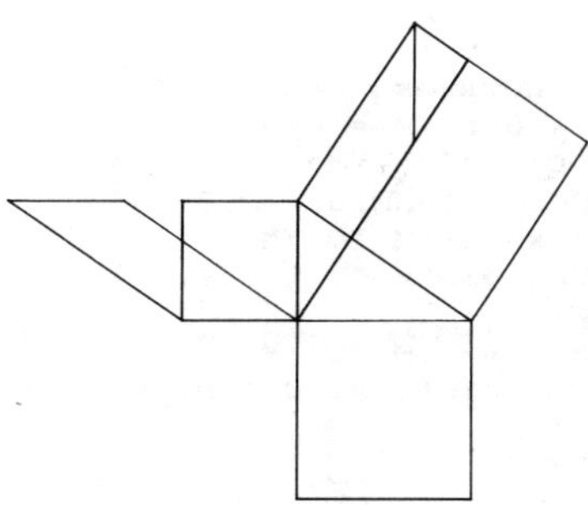

Fig. 5.9 — Output from Prog. 5.8.

5.5 BUILDING A TRANSFORMATION GEOMETRY PROGRAM

Chapter 17 of Oldknow and Smith's *Learning Mathematics with Micros* describes a technique for building up a "package" of transformation geometry utilities that can be used for geometrical exploration. The x,y coordinates of the I-th vertex of a shape S are stored in an array (S(1,I),S(2,I)) and the user can select from a "menu" of options to transform the coordinate array by one of several transformations.

5.5.1 The main block

The main part of such a program might look like:

```
 10 REM Prog.5.9 - Transformation geometry
 49 :
 50 MODE 1 : PROCsetup
 60 OX = 0.5*SW : OY = 0.5*SH
 70 SX = 50 : SY = 50
 99 :
100 DIM S(2,11)
110 REPEAT
120   CLS
130   PROCaxes
140   PROCdefine
150   REPEAT
160     PROCmenu
170     IF Ch<5 THEN PROCtransform
180   UNTIL Ch=5
190   PRINT "Do you want to run the program again" : INPUT R$
200 UNTIL R$ = "N" OR R$ = "n" OR R$ = "NO"
210 PRINT "Bye"
490 END
499 :
```

Here Ch will hold a number code representing the choice from the menu (e.g. Ch=1 might mean "translate"). Ch=5 is assumed to be the code for a choice to end.

5.5.2 Drawing axes

```
1000 DEF PROCaxes
1010   XM = 6 : YM = 4
1020   PROCjoin(0,-YM, 0,YM, 2)
1030   PROCjoin(-XM,0, XM,0, 2)
1040   FOR x = -XM TO XM
1050     PROCjoin(x,0, x,0.1, 2)
1060   NEXT x
1070   FOR y = -YM TO YM
1080     PROCjoin(0,y, 0.1,y, 2)
1090   NEXT y
1190 ENDPROC
1199 :
```

5.5.3 Defining a shape

PROCdefine allows users to define their own shape to be transformed. We have arbitrarily chosen a maximum of ten points on a scale of $-10<x<10$ and $-10<y<10$. The shape is drawn as the coordinates are entered. In order for the shape to "close-up" the last point is made coincident with the first. Again the choice of an input of 100 for x as being the signal to conclude the procedure is made arbitrarily. This procedure illustrates some crude principles of "idiot-proofing":

```
1200 DEF PROCdefine
1210   N = 1
1220   PRINT TAB(0,0);"Enter up to 10 pairs of coordinates x,y"
1230   PRINT "between -10 and 10. Use 100,100 to end"
1240   REPEAT
1250     PRINT "Point #";N;": ";
1260     INPUT " " X,Y
1270     Legal = (ABS(X)<=10) AND (ABS(Y)<=10)
1280     IF Legal AND N=1 THEN X1 = X : Y1 = Y
1290     IF Legal THEN PROCjoin(X1,Y1, X,Y, 3) : S(1,N)=X : S(2,N)=Y
1300     IF Legal THEN X1 = X : Y1 = Y : N = N + 1
1310   UNTIL N>10 OR X=100
1320   S(1,N) = S(1,1) : S(2,N) = S(2,1)
1330   PROCjoin(X1,Y1, S(1,1),S(2,1), 3)
1340   CLS : PROCaxes : PROCdisplay(3)
1390 ENDPROC
1399 :
```

5.5.4 Drawing the shape

All that we need to do to draw the shape is to select a colour, move to the first point and repeatedly draw the sides:

```
3000 DEF PROCdisplay(PC)
3010   X1 = S(1,1) : Y1 = S(2,1)
3020   FOR P = 2 TO N
3030     X = S(1,P) : Y = S(2,P)
3040     PROCjoin(X1,Y1, X,Y, PC)
3050     X1 = X : Y1 = Y
3060   NEXT P
3090 ENDPROC
3099 :
```

5.5.5 Selecting from a menu

The job of PROCmenu is to present a menu on the screen and to return a code Ch to the main program showing which choice has been made. In order to restrict printing to an area at the top of the screen the TAB(column, row) function is used, and the top NL lines of the screen are cleared by

PROCclear(NL):

```
1400 DEF PROCmenu
1410   PROCclear(3)
1420   REPEAT
1430     PRINT TAB(0,0);"1:translate, 2:reflect, 3:rotate,"
1440     PRINT "4:enlarge, 5:end   :";
1450     INPUT " " Ch : Ch = INT(Ch)
1460   UNTIL Ch>0 AND Ch<6
1490 ENDPROC
1499 :
3100 DEF PROCclear(NL)
3110   B$="" : S$ = " "
3120   FOR P = 1 TO 39
3130     B$ = B$ + S$
3140   NEXT P
3150   FOR P = 1 TO NL
3160     PRINT TAB(0,P-1);B$
3170   NEXT P
3190 ENDPROC
3199:
1215   PROCclear(3)
```

The main program "calls" another procedure, PROCtransform, which acts as a signal-box to steer us to the right part of the program for the chosen transformation:

```
1500 DEF PROCtransform
1510   IF Ch=1 THEN PROCtranslate  : ENDPROC
1520   IF Ch=2 THEN PROCreflect    : ENDPROC
1530   IF Ch=3 THEN PROCrotate     : ENDPROC
1540   IF Ch=4 THEN PROCenlarge    : ENDPROC
1590 ENDPROC
1599 :
```

5.5.6 Translating

The initial shape was drawn in white. Having selected a translation we need to collect some parameters which will be INPUT to the program. When these are gathered the shape is redrawn in red, the values of the coordinate array S are transformed and the new position of the shape is drawn in white:

```
1600 DEF PROCtranslate
1605   PROCclear(3)
1610   PRINT TAB(0,0);"Translate by x,y :";
1615   INPUT " " x,y
1620   PROCdisplay(1)
1625   PROCtran(x,y)
1630   PROCdisplay(3)
1640 ENDPROC
1649 :
```

The transformation of S is done in PROCtran:

```
1650 DEF PROCtran(x,y)
1655   FOR P = 1 TO N
1660     S(1,P) = S(1,P) + x
1665     S(2,P) = S(2,P) + y
1670   NEXT P
```

```
1690 ENDPROC
1699 :
```

5.5.7 Reflecting

The general structure of this section uses a mirror line of the form:

$$y = C + x*TAN(Th)$$

but we need to avoid the use of Th=90 degrees to specify a mirror line parallel to the y-axis. Thus this case has to be handled separately:

```
1700 DEF PROCreflect
1705  PROCclear(3)
1710  PRINT TAB(0,0);"Reflect : is the mirror-line parallel"
1720  PRINT "to the y-axis (Y/N)";
1730  INPUT R$
1740  IF R$ = "Y" OR R$ = "y" THEN PROCyReflect : ENDPROC
1745  PROCclear(3)
1750  PRINT TAB(0,0);"Enter the y-intercept and angle for"
1760  PRINT "the mirror-line :";
1770  INPUT " " C,Th
1775  PROCdisplay(1)
1780  PROCref(C,Th)
1785  PROCdisplay(3)
1790 ENDPROC
1799 :
```

The general reflection transformation is defined by:

```
1800 DEF PROCref(C,Th)
1810  Co = COS(Th*PI/180) : Si = SIN(Th*PI/180)
1820  FOR P = 1 TO N
1830   x = S(1,P) : y = S(2,P) - C
1840   S(1,P) = x*Co + y*Si
1850   S(2,P) = x*Si - y*Co + C
1860  NEXT P
1890 ENDPROC
1899 :
```

We still have the "special case" to consider:

```
1900 DEF PROCyReflect
1905  PROCclear(3) :
1906  PRINT TAB(0,0);"Reflection parallel to the y-axis."
1910  PRINT "Enter the x-intercept of the"
1914  PRINT "mirror-line :";
1915  INPUT " " D
1920  PROCdisplay(1)
1925  PROCyref(D)
1930  PROCdisplay(3)
1940 ENDPROC
1949 :
1950 DEF PROCyref(D)
1955  FOR P = 1 TO N
1960   S(1,P) = -S(1,P) + 2*D
1965  NEXT P
1990 ENDPROC
1999 :
```

5.5.8 Rotating

By now the pattern of the procedure definitions should be familiar. We need to collect information about the centre and angle of the rotation and use it to transform S:

```
2000 DEF PROCrotate
2005   PROCclear(3)
2010   PRINT TAB(0,0);"Rotation: enter the x,y coords of the"
2020   PRINT "centre and the angle of rotation:"
2030   INPUT " " Cx,Cy,Th
2040   PROCdisplay(1)
2050   PROCrot(Cx,Cy,Th)
2060   PROCdisplay(3)
2090 ENDPROC
2099 :
2100 DEF PROCrot(Cx,Cy,Th)
2110   Co = COS(Th*PI/180) : Si = SIN(Th*PI/180)
2120   FOR P = 1 TO N
2130     x = S(1,P) - Cx : y = S(2,P) - Cy
2140     S(1,P) = x*Co - y*Si + Cx
2150     S(2,P) = x*Si + y*Co + Cy
2160   NEXT P
2170 ENDPROC
2199 :
```

5.5.9 Enlarging

The only other transformation “on the menu” is that of an enlargement:

```
2200 DEF PROCenlarge
2205   PROCclear(3)
2210   PRINT TAB(0,0);"Enlargement: enter the x,y coords"
2220   PRINT "of the centre and the scale factor:"
2230   INPUT " " Cx,Cy,Sf
2240   PROCdisplay(1)
2250   PROCenl(Cx,Cy,Sf)
2260   PROCdisplay(3)
2270 ENDPROC
2299 :
2300 DEF PROCenl(Cx,Cy,Sf)
2310   FOR P = 1 TO N
2320     S(1,P) = (S(1,P)-Cx)*Sf + Cx
2330     S(2,P) = (S(2,P)-Cy)*Sf + Cy
2340   NEXT P
2350 ENDPROC
2399 :
```

5.5.10 A choice of types of use

As an alternative to providing an open framework for exploring geometric transformations we might start with displaying the intitial position of some predefined shape, then perform some “random” transformation(s) on it, displaying the result and invite the user to try to map the initial position onto the final position by a sequence of transformations. If we change line 140 to:

```
140  PROCchoice
```

and line 100 to:

```
100 DIM S(2,11),T(2,11)
```

then we can offer a choice of the "electronic-blackboard" mode or the "problem-solving" mode of use:

```
4200 DEF PROCchoice
4210  REPEAT
4215   PROCclear(3)
4220   PRINT TAB(0,0);"Do you want to: 1) explore on your own"
4230   PRINT "or 2) solve a problem? Type 1 or 2:";
4240   INPUT " " Ch : Ch = INT(Ch)
4250  UNTIL Ch>0 AND Ch<3
4255  IF Ch<>2 THEN PROCdefine : ENDPROC
4260  PROCclear(3)
4265  PRINT TAB(0,0);"Map the white shape to the"
4270  PRINT "yellow one. ";
4275  PROCgenerate
4280  PRINT "Press any key to start: ";
4285  INPUT G$
4290 ENDPROC
```

5.5.11 Choosing a random transformation

Finally we need to define PROCgenerate to define a shape and pick a random transformation. RND(4) will pick "at random" one of the numbers 1, 2, 3 or 4. Thus we use this to make a random choice of which of our four transformations "on the menu" will be used. Depending on the transformation chosen we will also need to select the values of one or more parameters at random. Lines 4010–4050 define a small flag that is stored in both the arrays S and T. The "target" shape is drawn in yellow and then the array S is "refilled" from T to draw the original shape in white:

```
4000 DEF PROCgenerate
4010  N = 4 : RESTORE 4050
4020  FOR P = 1 TO N
4030   READ S(1,P), S(2,P)
4035   T(1,P) = S(1,P) : T(2,P) = S(2,P)
4040  NEXT P
4050  DATA 1,1, 1,3, 2,2, 1,2
4060  Ch = RND(4)
4070  IF Ch=1 THEN PROCtran(RND(7)-4,RND(9)-5)
4080  IF Ch=2 THEN PROCref(RND(7)-4,RND(8)*45)
4090  IF Ch=3 THEN PROCrot(RND(7)-4,RND(7)-4,RND(8)*45)
4100  IF Ch=4 THEN PROCenl(RND(7)-4,RND(7)-4,RND(4)-2.5)
4110  PROCdisplay(2)
4120  FOR P = 1 TO N
4130   S(1,P) = T(1,P)
4140   S(2,P) = T(2,P)
4150  NEXT P
4160  PROCdisplay(3)
4170 ENDPROC
4199 :
```

5.5.12 Extensions

Although the resulting program is really rather long and, eventually, fairly sophisticated it has been constructed from "bite-size" chunks. If you are

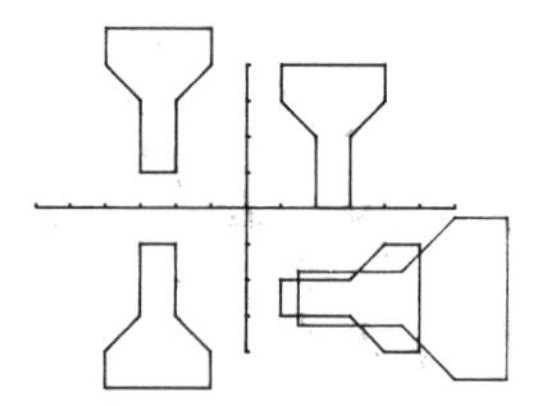

Fig. 5.10 — Some output from transformation geometry package.

interested in constructing reasonably complicated programs for yourself then it might be an idea to see if you can gain confidence in "tacking bits on" to this package. For example, you could add a shear and/or one-way stretch to the transformation menu. You could select a random combination of two or more transformations in PROCgenerate, or use a more involved shape. You could include some more sophisticated error-trapping to make the program more robust.

5.6 EIGENVALUES AND EIGENVECTORS

In this section we establish some tools for the examination of eigenvectors and eigenvalues of 2×2 matrices that can be extended to higher order matrices.

An n×n matrix M has an eigenvalue k and associated eigenvector **e** if:

$$M.\mathbf{e} = k.\mathbf{e}$$

There are direct algebraic and numerical methods of finding the eigenvalues from the solution of the n-th order polynomial equation that is derived from the associated "characteristic equation":

$$(M - k.I) = 0$$

where I is the n×n identity matrix. However, we shall take a different approach to hunt for the "dominant" eigenvalue (if one exists).

If a matrix has at least one real eigenvalue (and associated eigenvector) then one of its eigenvalues is said to be dominant if its absolute value is numerically larger than that of each of the others (i.e. ignoring minus signs).

In this case we can start with an initial vector **vo** and repeatedly multiply it by the matrix M:

$$\mathbf{v1} = M.\mathbf{v0} \qquad \mathbf{v2} = M.\mathbf{v1} \qquad \mathbf{v3} = M.\mathbf{v2} \qquad \ldots$$

and the sequence of vectors **vi** tends to a limiting vector **E** which is the eigenvector associated with the dominant eigenvalue L1. If the dominant eigenvalue has an absolute value greater than one then the sequence of vectors, although tending to a fixed direction, will get larger and larger in length. In order both to achieve information about the size of the dominant eigenvalue (if it exists!) and in order to represent the process graphically it is preferable to modify the process slightly.

The length (or "norm") of a vector is usually defined as the square-root of the sum of squares of its components — this is a generalisation of Pythagoras. Thus for any vector **w** we can find an associated unit vector **v** by first calculating the length l of **w** and making each component of **v** be the corresponding component of **w** divided by l. This process is called "normalising" the vector **w**. We shall denote the length of the vector **v** by $|\mathbf{v}|$.

If we now start with v0 as a unit vector (i.e. of length 1) we can repeatedly multiply by M and then normalise to achieve a new sequence of vectors, all of length 1:

$$\mathbf{w1} = M.v0 \qquad v1 = \mathbf{w1}/|\mathbf{w1}|$$
$$\mathbf{w2} = M.v1 \qquad v2 = \mathbf{w2}/|\mathbf{w1}|$$
$$\mathbf{w3} = M.v3 \qquad \ldots$$

The limiting value **E** (if there is one) of the sequence of vectors vi is the unit eigenvector corresponding to the dominant eigenvalue L1 (if there is one) of M which is the limiting value of the sequence $|\mathbf{wi}|$.

We can now turn this process into a program quite easily. The REPEATed process of (a) multiplying **v** by M to get **w**, (b) displaying the vectors and (c) normalising **w** to get **v** needs to be terminated by some condition. This will be when two successive values of $|\mathbf{w}|$ are close enough for "all practical purposes". The following program performs the process:

```
 10 REM Prog.5.10 - Eigenvectors of a 2x2 matrix
 49 :
 50 MODE 1 : PROCsetup
 60 OX = 0.5*SW : OY = 0.5*SH
 70 SX = 100 : SY = 100
 99 :
100 PROCaxes
110 PROCdefine
120 REPEAT
130   PROCmultiply
140   PROCdisplay
150   PROCnormalise
160 UNTIL test<delta
170 PRINT L
490 END
499 :
```

PROCaxes is, by now, a familiar procedure to set up the screen. PROCdefine will need to fill the matrix M and to specify a starting value for the vector **V** (of length 1). It also needs to define the value of "delta"used to tell when two successive values of $L=|\mathbf{W}|$ are sufficiently close.

PROCmultiply simply defines multiplication of a 2×2 matrix and a 2×1 column vector.

PROCdisplay just puts the current positions of both **V** and **W** on the screen.

PROCnormalise calculates the length L of **W** and uses it to generate the corresponding normalised vector **V**. The value of "test" is the difference between the current value L and its previous value LO:

```
1500 DEF PROCaxes
1510   PROCjoin(-5,0, 5,0, 1)
1520   PROCjoin(0,-5, 0,5, 1)
1530   PRINT'';"Eigenvalue";TAB(15);"Vector-x";TAB(28);"Vector-y"
1590 ENDPROC
1599 :
1600 DEF PROCdefine
1610   DIM M(2,2), V(2), W(2), E(2), F(2)
1620   A = 3 : B = 2 : C = 1 : D = 2
1630   M(1,1) = A : M(1,2) = B
1640   M(2,1) = C : M(2,2) = D
1650   V(1) = 1/SQR(2) : V(2) = 1/SQR(2)
1660   LO = 1 : delta = 0.00005
1690 ENDPROC
1699 :
1700 DEF PROCmultiply
1710   W(1) = M(1,1)*V(1) + M(1,2)*V(2)
1720   W(2) = M(2,1)*V(1) + M(2,2)*V(2)
1790 ENDPROC
1799 :
1800 DEF PROCdisplay
1810   PROCjoin(0,0, V(1),V(2), 2)
1820   PROCjoin(0,0, W(1),W(2), 3)
1890 ENDPROC
1899 :
1900 DEF PROCnormalise
1910   L = SQR(W(1)^2+W(2)^2)
1920   V(1)=W(1)/L : V(2)=W(2)/L
1930   test = ABS(L-LO) : LO = L
1990 ENDPROC
1999 :
```

At the end of this program L holds the value of the dominant eigenvector (if there is one). The matrix used in the program is:

$$M = \begin{bmatrix} 3 & 2 \\ 1 & 2 \end{bmatrix}$$

Try other initial unit test vectors such as V(1)=1, V(2)=0 or V(1)=0, V(2)=1. Are the limiting values of L and **V** the same for all choices of the initial vector **V**?

Try other 2×2 matrices and see what happens in the cases when either(a) there are no real eigenvalues (such as for a rotation matrix) or (b) there are two real eigenvalues of equal magnitude.

If the process has worked properly and the matrix does have a dominant eigenvalue L1 then it should be relatively easy to separate out the other eigenvalue L2. (Remember that a 2x2 matrix yields a quadratic equation for the eigenvalues, and thus they are either both real — but possibly equal — or both complex.

If the matrix M has a dominant eigenvalue L1 with unit eigenvector **E**, and another eigenvalue L2 with unit eigenvector **F** we can use **E** and **F** as a new "basis" for vectors in the plane. Thus our initial search vector **V** can be expressed as a linear combination of **E** and **F**:

$$\mathbf{V} = a.\mathbf{E}+b.\mathbf{F}$$

If we now multiply **V** by the matrix M:

$$M.\mathbf{V} = M.(a.\mathbf{E}+b.\mathbf{F}) = a.M.\mathbf{E}+b.M.\mathbf{F} = a.L1.\mathbf{E}+b.L2.\mathbf{F}$$

So if we now subtract L1.**V** we have:

$$M.\mathbf{V} - L1.V = b.(L2{-}L1).\mathbf{F} = \mathbf{W} \qquad \text{say}$$

If we normalise **W** to obtain a unit vector then this is the eigenvector **F**. If we now multiply by M again:

$$M.\mathbf{F} = L2.\mathbf{F}$$

and hence the eigenvalue L2 is the length of M.**F**
We just need to add two lines to the program:

```
180 PROCseparate
190 PROCeigenvectors
```

and define the two new procedures:

```
1000 DEF PROCseparate
1010   L1 = L : E(1) = V(1) : E(2) = V(2)
1020   PRINT';L1;TAB(13);E(1);TAB(26);E(2)
1030   V(1) = 1/SQR(2) : V(2) = 1/SQR(2)
1040   PROCmultiply
1050   W(1) = W(1) - L1*V(1)
1060   W(2) = W(2) - L1*V(2)
1070   PROCnormalise
1080   PROCmultiply
1090   PROCnormalise
1100   L2 = L : F(1) = V(1) : F(2) = V(2)
1110   PRINT';L2;TAB(13);F(1);TAB(26);F(2)
1120 ENDPROC
1199 :
1200 DEF PROCeigenvectors
1210   PROCjoin(0,0, L1*E(1),L1*E(2), 1)
1220   PROCjoin(0,0, L2*F(1),L2*F(2), 1)
1290 ENDPROC
1299 :
```

The final display shows the pair of eigenvectors in proportion to their eigenvalues.

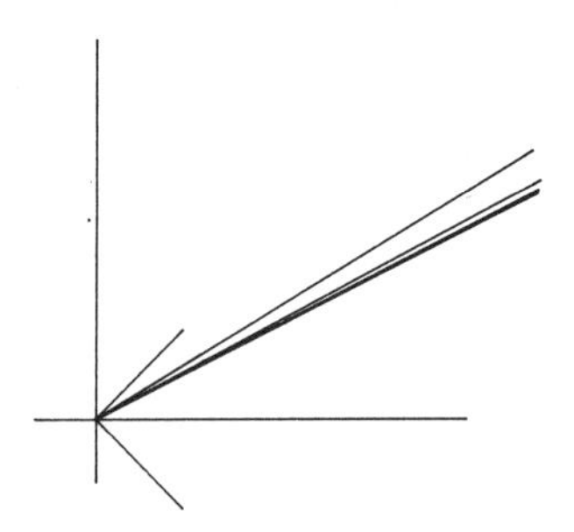

Fig. 5.11 — Output from Prog. 5.10.

5.7 MATRIX RECURRENCE

We have met the idea of generating a circle of radius R, centre the origin by defining an initial vector by x=R, y=0 and repeatedly transforming this vector by the matrix for a rotation through a small angle Th. Now, on some computer systems much saving in time and space can be achieved if it is possible to avoid any computation of the trigonomeric functions SIN, COS etc.

We can write the general matrix recurrence that transforms the point (X,Y) to the point (XT,YT) as:

$$XT = A*X + B*Y$$

$$YT = C*X + D*Y$$

For a rotation we used A = D = COS(Th), C = SIN(Th) and B = −SIN(Th) but, if Th is small, we could use, say, some terms from the MacLaurin series for SIN and COS such as:

$$\text{SIN(Th)} = \text{Th} \qquad \text{COS(Th)} = 1 - (\text{Th}^2)/2$$

Now the determinant **det** of the transformation matrix is det = A*D−B*C. For a rotation matrix this gives

$$\text{det} = \text{COS}^2\text{(Th)} + \text{SIN}^2\text{(Th)} = 1$$

but, for the approximate case, we have

$$\text{det} = A*D-B*C = (1 - (\text{Th}^2)/2)^2 + \text{Th}^2 = 1 + (\text{Th}^4)/4$$

and, if we use this to generate a curve, we find that it does not join up. In order to keep the value of the determinant at 1 we can "shuffle" the approximations to give:

$$\begin{bmatrix} 1 & -\text{Th} \\ \text{Th} & 1-\text{Th}^2 \end{bmatrix}$$

The following program uses this recurrence:

```
 10 REM Prog.5.11 - Curves generated by matrix recurrence
 49 :
 50 MODE 1 : PROCsetup
 60 OX = 0.5*SW : OY = 0.5*SH
 70 SX = 100 : SY = 100
 99 :
100 X1 = 4 : Y1 = 0
110 Th = 0.2
120 A = 1 : B = -Th : C = Th : D = 1-Th*Th
130 REPEAT
140  X = A*X1 + B*Y1
150  Y = C*X1 + D*Y1
160  PROCjoin(X,Y, X1,Y1, FC)
170  X1 = X : Y1 = Y
180 UNTIL "Cows" = "Home"
490 END
```

Note that line 180 means that the program continues forever (or at least until the ESCAPE key is pressed).

For small values of Th the curve is a good approximation to a circle.

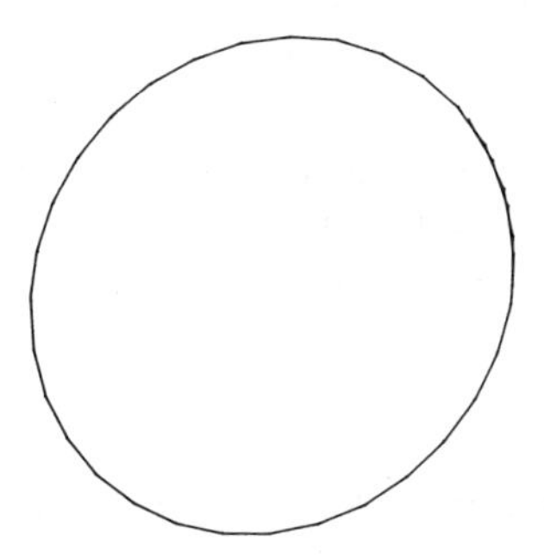

Fig. 5.12 — Output from Prog. 5.11.

How could you make it stop when it closed up?

For larger values of Th the curve is clearly an ellipse with its major axis along the y = x and the eccentricity seems to be a function of Th. See if you can prove that the eccentricity **e** of this ellipse is given by:

e = SQR((2*Th)/(2+Th))

6

Splines, approximations and curves in space

In this chapter we extend the idea of a point on a line AB as the weighted average of the coordinates of A and B to produce curves as the weighted averages of a set of points. Since four or more points need not lie in the same plane then such curves may well be twisted curves in space and so we shall need to develop some techniques for viewing these. We also take a little diversion to see how we can use some of the techniques developed so far to explore ideas of Fourier and Chebychev series.

6.1 BEZIER CURVES

We have frequently made use of the vector equation of a line AB in the form:

$$\mathbf{r} = (1-t)*\mathbf{a} + t*\mathbf{b}$$

where t is some parameter, to obtain the coordinates (Px,Py) of a point P on the line joining A(Ax,Ay) to B(Bx,By):

$$x = (1-t)*Ax + t*Bx \qquad y = (1-t)*Ay + t*By$$

To get a feel for this try putting it into a program and fiddling with the value of t:

```
 10 REM Prog.6.1 - Vector equation of a line
 49 :
 50 MODE 1 : PROCsetup
 60 OX = 0.5*SW : OY = 0.5*SH
 70 SX = 100 : SY = 100
 99 :
100 Ax = -2 : Ay = 0 : PROCdot(Ax,Ay,FC)
110 Bx = 4 : By = 4 : PROCdot(Bx,By,FC)
120 REPEAT
```

```
130   INPUT t
140   x = (1-t)*Ax + t*Bx
150   y = (1-t)*Ay + t*By
160   PROCdot(x,y, FC)
170 UNTIL t>20
490 END
499 :
500 DEF PROCsetup
510   SW = 1280 : SH = 1024
520   NC = 3 : FC = 3
590 ENDPROC
599 :
600 DEF PROCdot(X,Y,PC)
610   GCOL 0,PC
620   PLOT 69, OX + X*SX, OY + Y*SY
690 ENDPROC
699 :
700 DEF PROCjoin(X1,Y1,X,Y,PC)
710   GCOL 0,PC
720   MOVE OX + X1*SX, OY + Y1*SY
730   DRAW OX + X*SX, OY + Y*SY
790 ENDPROC
799 :
```

This is rather a crude program but you can explore how the value of t determines the position of the point in relation to A and B. A simple extension is provided by adding:

```
130 FOR t = 0 TO 1 STEP 1/16
170 NEXT t
```

What happens to the points on the line as the step varies? Try using this idea together with the **dot** and **join** commands to make up various sorts of dashed, broken and dotted lines.

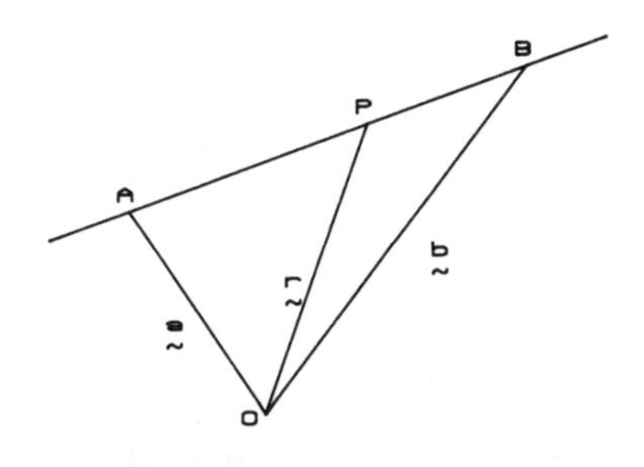

Fig. 6.1.

Lines 140,150 form (X,Y) as a **weighted average** of (Ax,Ay) and (Bx,By) — of course there is no restriction to the vectors being only in two-dimensions but it does make display on the screen easier.

We could explore the idea of (X,Y) as being a weighted average of three or more points. As the **weights**, (1−t) and t, are linear in t it might be reasonable to expect that, for three points, they could be quadratic in t. A nice source of weights comes from Pascal's Triangle (or the binomial coefficients). For three points these are:

$$(1-t)^2 \quad 2(1-t)t \quad t^2$$

We can easily extend the idea of Prog. 6.1 to draw a curve generated by three points using these weighting functions. The coordinates of the points, A,B,C have been 'renamed' (Xa,Ya), (Xb,Yb), (Xc,Yc) and the triangle connecting them is first drawn in red:

```
 10 REM Prog.6.2 - Bezier curve on 3 points
 49 :
 50 MODE 1 : PROCsetup
 60 OX = 0.5*SW : OY = 0.5*SH
 70 SX = 100 : SY = 100
 99 :
100 Xa = -4 : Ya = -4 : PROCdot(Xa,Ya,1)
110 Xb = -1 : Yb = 4 : PROCjoin(Xa,Ya, Xb,Yb, 1)
120 Xc = 3 : Yc = 3 : PROCjoin(Xb,Yb, Xc,Yc, 1)
130 PROCjoin(Xc,Yc, Xa,Ya, 1)
140 X1 = Xa : Y1 = Ya
150 FOR t = 0 TO 1 STEP 1/16
160  u = 1-t : f = u*u
170  g = 2*u*t : h = t*t
180  X = f*Xa + g*Xb + h*Xc
190  Y = f*Ya + g*Yb + h*Yc
200  PROCjoin(X1,Y1, X,Y, 3)
210  X1 = X : Y1 = Y
220 NEXT t
490 END
```

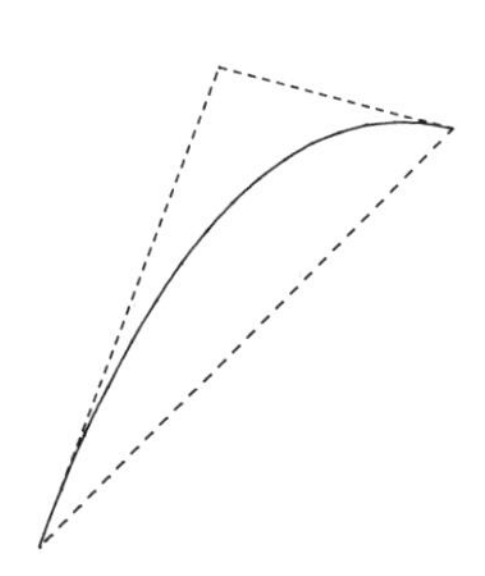

Fig. 6.2.

Can you prove that the curve is tangent to AB at A and to BC at C? What is the geometric significance of the point given by t=0.5? We might conjecture that the tangent to the curve at t=0.5 is parallel to AC. The following additions draw in a median of the triangle from B to AC and draw a line segment parallel to AC through its midpoint:

```
230 :
300 PROCjoin(Xb,Yb, (Xa+Xc)/2,(Ya+Yc)/2, 1)
310 dx = (Xc-Xa)/5 : dy = (Yc-Ya)/5
320 x = (Xa + 2*Xb + Xc)/4
330 y = (Ya + 2*Yb + Yc)/4
340 PROCjoin(x-dx,y-dy, x+dx,y+dy, 2)
```

Adapt the program to explore the effect on the curve of keeping A and C fixed and moving B around. Try extending the idea to four or more points.

The principle used is that for three points a curve is generated by a vector equation of the form:

$$\mathbf{r} = f(t)*\mathbf{a}+g(t)*\mathbf{b}+h(t)*\mathbf{c}$$

where the functions f(t), g(t) and h(t) are **weighting functions**. That is, for $0 \leqslant t \leqslant 1$:

$$f(t) \geqslant 0 \quad g(t) \geqslant 0 \quad h(t) \geqslant 0 \quad \text{and} \quad f(t) + g(t) + h(t) = 1$$

Bezier curves use the terms of the binomial series:

$$C(n,r,t) = \frac{n!}{r!(n-r)!} . t^r . (1-t)^{n-r}$$

as the weighting functions. There are other common choices which give different spline representations, such as B-spline and Hermite, that we shall meet later.

The additions to Prog. 6.2 to make a curve on four points are straightforward:

```
125 Xd = 4 : Yd = -4 : PROCjoin(Xc,Yc, Xd,Yd, 1)
130 PROCjoin(Xd,Yd, Xa,Ya, 1)
160  u = 1-t : f = u*u*u : g = 3*u*u*t
170  h = 3*u*t*t : i = t*t*t
180  X = f*Xa + g*Xb + h*Xc + i*Xd
190  Y = f*Ya + g*Yb + h*Yc + i*Yd
```

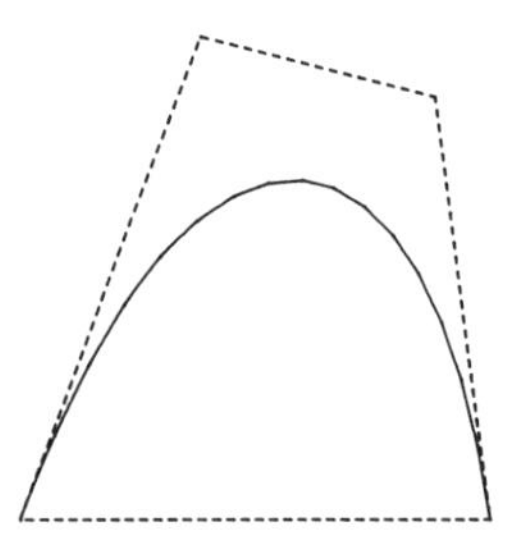

Fig. 6.3.

These curves were applied by P. Bezier of Renault cars for use in computer-aided design (CAD) in the 1960s and are also one of a class of Bernstein polynomials.

To generalise beyond four points we really need to use arrays to store the coordinates (X(i),Y(i)) of the i-th point and loops to perform the summations:

$$X = \sum_{r} C(n,r,t)*X(r) \qquad Y = \sum_{r} C(n,r,t)*Y(r)$$

In fact the binomial coefficients are most easily calculated by observing that:

$$\frac{C(n,r,t)}{C(n,r-1,t)} = \frac{n+1-r}{r}.\frac{t}{1-t}$$

In the following program there are n points, labelled from 0 to m (where m = N − 1) and c holds the current value of the binomial coefficient nCr:

```
 10 REM Prog.6.3 - Bezier curve on n points
 49 :
 50 MODE 1 : PROCsetup
 60 OX = 0.1*SW : OY = 0.1*SH
 70 SX = 100 : SY = 100
 99 :
100 n = 5 : m = n-1
110 DIM X(m),Y(m)
120 READ X1,Y1 : X(0) = X1 : Y(0) = Y1
130 FOR i = 1 TO m
140   READ X,Y
150   PROCjoin(X1,Y1, X,Y, 3)
160   X(i) = X : Y(i) = Y : X1 = X : Y1 = Y
170 NEXT i
180 DATA 1,1, 4,6, 7,2, 1,8, 6,8
190 :
200 s = 1/32 : X1 = X(0) : Y1 = Y(0)
210 FOR t = 0 TO 1-s STEP s
220   u = 1-t : c = u^m : r = t/u
```

```
230   X = c*X(0) : Y = c*Y(0)
240   FOR i = 1 TO m
250    c = c*r*(m+1-i)/i
260    X = X + c*X(i)
270    Y = Y + c*Y(i)
280   NEXT i
290   PROCjoin(X1,Y1, X,Y, 3)
300   X1 = X : Y1 = Y
310 NEXT t
320 X = X(m) : Y = Y(m)
330 PROCjoin(X1,Y1, X,Y ,3)
490 END
```

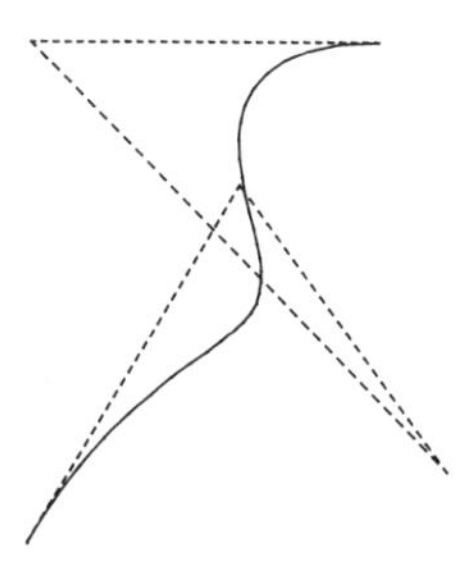

Fig. 6.4.

Clearly there is no reason why we should not include a z-component and use 3D coordinates:

```
275    Z = Z + c*Z(i)
```

Later we shall develop some techniques for displaying such space curves.

6.2 B-SPLINE CURVES

Another set of weighting functions in common use gives a family of curves called the B-spline curves. For two points we would expect the linear weighting functions: $(1-t)$ and t to duplicate those of the straight-line (and Bezier). For three points we have:

$$f(t) = (1-t)^2/2 \qquad \text{and} \qquad h(t) = t^2/2$$

and, using the property of weights that $f+g+h = 1$ gives:

$$g(t) = t(1-t)+\tfrac{1}{2}$$

The changes to Prog. 6.2 are very slight:

```
 10 REM Prog.6.4 - B-spline curve on 3 points
 49 :
 50 MODE 1 : PROCsetup
 60 OX = 0.5*SW : OY = 0.5*SH
 70 SX = 100 : SY = 100
 99 :
100 Xa = -4 : Ya = -4 : PROCdot(Xa,Ya,1)
110 Xb = -1 : Yb = 4 : PROCjoin(Xa,Ya, Xb,Yb, 1)
120 Xc = 3 : Yc = 3 : PROCjoin(Xb,Yb, Xc,Yc, 1)
130 PROCjoin(Xc,Yc, Xa,Ya, 1)
140 X1 = (Xa+Xb)/2 : Y1 = (Ya+Yb)/2
150 FOR t = 0 TO 1 STEP 1/16
160  f = (1-t)^2/2 : g = 0.5 + t*(1-t)
170  h = t^2/2
180  X = f*Xa + g*Xb + h*Xc
190  Y = f*Ya + g*Yb + h*Yc
200  PROCjoin(X1,Y1, X,Y, 3)
210  X1 = X : Y1 = Y
220 NEXT t
490 END
```

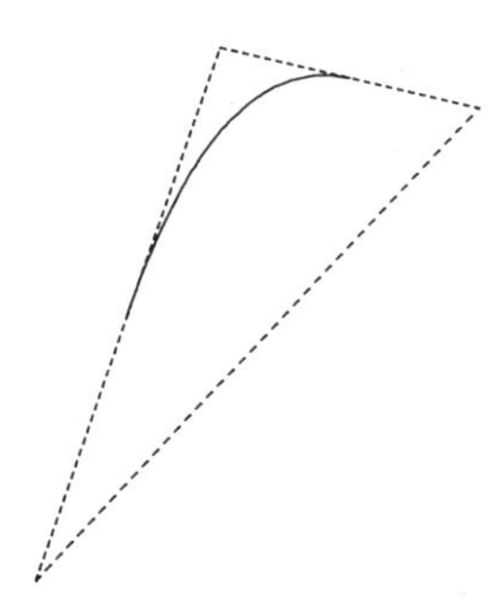

Fig. 6.5.

Note that the curve does not now pass through any of the three points for the parameter range $0 \leqslant t \leqslant 1$ and hence the need for the change to line 140. What happens if t goes outside that range? Are there any obvious geometrical properties of the curve in relation to the three defining points?

For four points the weighting functions are even stranger. Generalising we might expect $f(t) = (1-t)^3/6$ and $i(t) = t^3/6$ but the forms of g(t) and h(t) are rather strange:

$$g(t) = ((2-t)^3 - 4*(1-t)^3)/6 \qquad h(t) = ((1+t)^3 - 4*t^3)/6$$

A derivation of these functions can be found in Oldknow and Smith, Section 14.3.

The following changes are all that are needed:

```
125 Xd = 4 : Yd = -4 : PROCjoin(Xc,Yc, Xd,Yd, 1)
130 PROCjoin(Xd,Yd, Xa,Ya, 1)
140 :
160  f = (1-t)^3/6 : g = ((2-t)^3 - 4*(1-t)^3)/6
170  h = ((1+t)^3 - 4*t^3)/6 : i = t^3/6
180  X = f*Xa + g*Xb + h*Xc + i*Xd
190  Y = f*Ya + g*Yb + h*Yc + i*Yd
195  IF t=0 THEN X1 = X : Y1 = Y
```

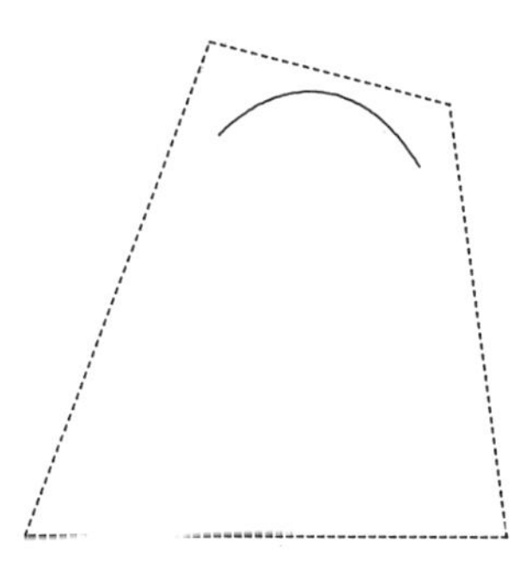

Fig. 6.6.

Now such **cubic** curve arcs generated by B-splines on sets of 4 points are frequently used as the basis of computer-aided techniques for representing smooth curves, and we shall explore these further shortly. However, for the sake of completeness, we can give a general formulation of the B-spline curve generated by N points. First we introduce a 'piece-wise exponential function' which raises x to the power p if x>0 and yields 0 otherwise:

DEF FNP(X,p) = −(X>0)*(X ↑ p)

this uses a mixture of logical and arithmetic operations. The first bracket compares X with 0 and returns a value of TRUE or FALSE depending upon the result. In most microcomputers TRUE is represented by the number −1 and FALSE by 0. Thus if X⩽0 then the first bracket is zero and the result is zero. If X>0 then the first bracket has value −1 and so we need to multiply this by −1 to keep the signs correct! This mixing of logical, relational and arithmetic operators gives a powerful way of denoting piecewise definitions.

We next observe that the basis cubic B-spline polynomial can be written as:

B(t) = (FNP(t+2,3)−4*FNP(t+1,3)+6*FNP(t,3)
−4*FNP(t−1,3)+FNP(t−2,3))/6

to see the general structure in terms of the binomial coefficients. A curve on n points will be of degree m, where m=n−1:

```
10 REM Prog.6.5 - B-spline curve on N points
49 :
50 MODE 1 : PROCsetup
60 OX = 0.1*SW : OY = 0.1*SH
```

```
  70 SX = 100 : SY = 100
  99 :
 100 n = 5 : m = n-1
 110 DIM X(m),Y(m)
 120 READ X1,Y1 : X(0) = X1 : Y(0) = Y1
 130 FOR i = 1 TO m
 140  READ X,Y
 150  PROCjoin(X1,Y1, X,Y, 3)
 160  X(i) = X : Y(i) = Y : X1 = X : Y1 = Y
 170 NEXT i
 180 DATA 1,1, 4,6, 7,2, 1,8, 6,8
 190 :
 200 s = 1/32
 210 FOR t = 0 TO 1 STEP s
 220  X = 0 : Y = 0
 230  FOR k = 0 TO m
 240   PROCB(k+1-t-n/2, m)
 250   X = X + B*X(k)
 260   Y = Y + B*Y(k)
 270  NEXT k
 280  IF t=0 THEN X1 = X : Y1 = Y
 290  PROCjoin(X1,Y1, X,Y, 3)
 300  X1 = X : Y1 = Y
 310 NEXT t
 490 END
 499 :
1000 DEF PROCB(p,L)
1005  IF L=0 THEN B = -((p<=0.5) AND (p>=-0.5)) : ENDPROC
1010  L1 = L + 1 : c = 1 : a = p + L1/2 : sum = FNP(a,L)
1015  FOR i = 0 TO L
1020   c = -c*(L1-i)/(i+1)
1025   a = a-1
1030   sum = sum + c*FNP(a,L)
1035  NEXT i
1040  FOR i = 1 TO L
1045   sum = sum/i
1050  NEXT i
1055  B = sum
1090 ENDPROC
1099 :
1100 DEF FNP(X,P) = -(X>0)*(X^P)
1199 :
```

Lines 1000,1090 are written as a procedure to evaluate B(p) for a given choice of L. In some versions of Basic (including BBC Basic) this could be rewritten as a multi-line function, in others it will have to be converted into a subroutine.

Now that last program is really rather cumbersome and slow. There are far more efficient algorithms for computing B-splines, many of which are recursive, that can be found in the technical literature (see, for example, Barnhill and Riesenfeld). However, for most applications we only need the cubic spline generated by four points, so it will be helpful to have this as a procedure:

```
1400 DEF PROCspline(Xa,Ya, Xb,Yb, Xc,Yc, Xd,Yd, PC)
1405  FOR t = 0 TO 1 STEP 1/32
1410   w4 = t*t*t
1415   s  = 1-t : w1 = s*s*s
1420   r  = 1+t : w3 = r*r*r - 4*w4
1425   w2 = 6-w4-w3-w1
```

```
1430    X   = (w1*Xa  + w2*Xb  + w3*Xc  + w4*Xd)/6
1435    Y   = (w1*Ya  + w2*Yb  + w3*Yc  + w4*Yd)/6
1440    IF t=0 THEN X1 = X : Y1 = Y
1445    PROCjoin(X1,Y1, X,Y, PC)
1450    X1 = X : Y1 = Y
1455   NEXT t
1490 ENDPROC
1499 :
```

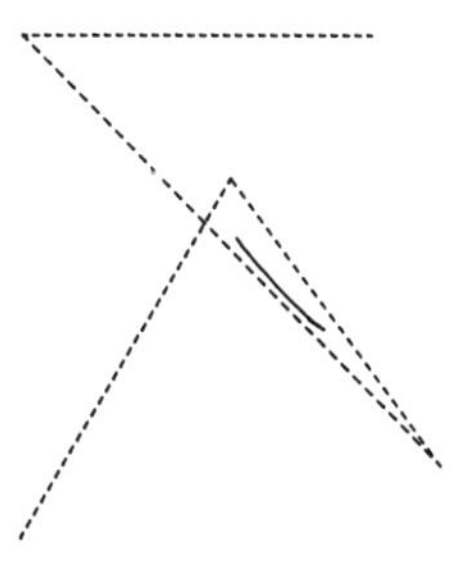

Fig. 6.7.

Suppose we have N points (N>4) P0, P1, P2, ..., Pm where m=N−1. We can generate cubic arcs from P0.P1.P2.P3, P1.P2.P3.P4, P2.P3.P4.P5, P3.P4.P5.P6 etc. Now the vital geometric property of such adjacent B-spline arcs is that they join each other smoothly, as can be seen from the following program:

```
 10 REM Prog.6.6 - An open cubic B-spline curve on N points
 50 MODE 1 : PROCsetup
 60 OX = 0.1*SW : OY = 0.1*SH
 70 SX = 100 : SY = 100
 99 :
100 n = 6 : m = n-1
110 DIM X(m),Y(m)
120 READ X1,Y1 : X(0) = X1 : Y(0) = Y1
130 FOR i = 1 TO m
140  READ X,Y
150  PROCjoin(X1,Y1, X,Y, 3)
160  X(i) = X : Y(i) = Y : X1 = X : Y1 = Y
170 NEXT i
180 DATA 1,1, 2,9, 8,7
190 DATA 5,5, 10,3, 5,1
200 :
230 FOR i = 0 TO m-3
240  PC = 3
250  Xa = X(i)   : Ya = Y(i)
260  Xb = X(i+1) : Yb = Y(i+1)
270  Xc = X(i+2) : Yc = Y(i+2)
280  Xd = X(i+3) : Yd = Y(i+3)
290  PROCspline(Xa,Ya, Xb,Yb, Xc,Yc, Xd,Yd, PC)
300 NEXT i
490 END
499 :
```

In order to make the curve extend to the end-points P0 and Pm we can start

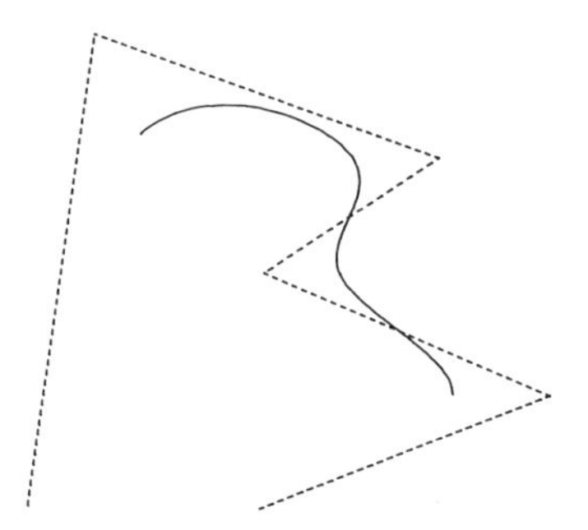

Fig. 6.8.

with some arcs made from coincident points: P0.P0.P0.P1, P0.P0.P1.P2, and finish similarly. In order to show the individual arcs more clearly we can change colours between arcs. The following additions perform this:

```
200 x0 = X(0) : y0 = Y(0)
210 PROCspline(x0,y0, x0,y0, x0,y0, X(1),Y(1), 3)
220 PROCspline(x0,y0, x0,y0, X(1),Y(1), X(2),Y(2), 2)

240  PC = 2 + (i MOD 2)

310 PC = 2 + (m MOD 2)
320 xm = X(m) : ym = Y(m) : xp = X(m-1) : yp = Y(m-1)
330 PROCspline(X(m-2),Y(m-2), xp,yp, xm,ym, xm,ym, PC)
340 PC = 2 + (n MOD 2)
350 PROCspline(xp,yp, xm,ym, xm,ym, xm,ym, PC)
```

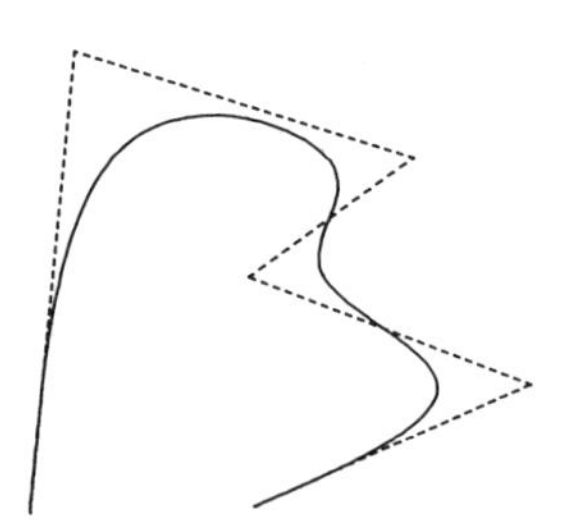

Fig. 6.9.

An alternative approach to this is to make the choice of successive sets of four points cyclic. Thus after Pm−3.Pm−2.Pm−1.Pm we use Pm−2.Pm−1.Pm.P0, Pm−1.Pm.P0.P1 and Pm.P0.P1.P2 to obtain a closed curve:

```
200 PROCjoin(X(m),Y(m), X(0),Y(0), 1)
210 :
220 :
230 FOR i = 0 TO m
240   PC = 2 + (i MOD 2)
250   Xa = X(i) : Ya = Y(i) : i2 = (i+1) MOD n
260   Xb = X(i2) : Yb = Y(i2) : i3 = (i2+1) MOD n
270   Xc = X(i3) : Yc = Y(i3) : i4 = (i3+1) MOD n
280   Xd = X(i4) : Yd = Y(i4)
290   PROCspline(Xa,Ya, Xb,Yb, Xc,Yc, Xd,Yd, PC)
300 NEXT i
310 :
320 :
330 :
340 :
350 :
```

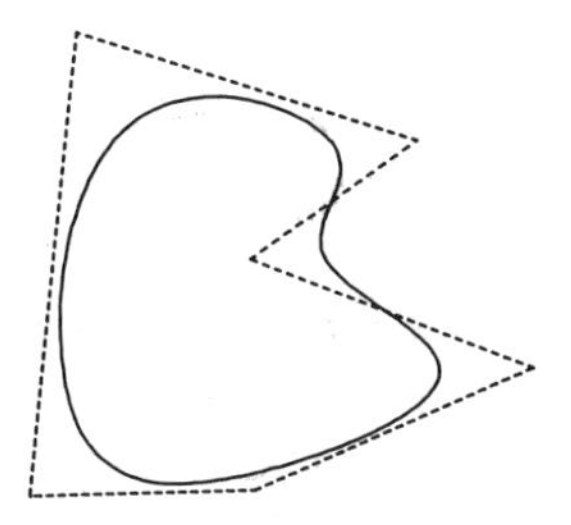

Fig. 6.10.

Both the Bezier curve on N points and the open cubic B-spline on N points share the property that they do not, in general, pass through any data points except the endpoints. The intermediate points are usually called the **control points**. For Bezier curves we have the property that P0.P1 defines the starting tangential direction and that Pm−1.Pm similarly defines the finishing direction. Are there similar geometric properties for B-splines?

Since the Bezier representation is an m-th degree polynomial then changing any one control point will alter the shape of the curve over its whole extent — a phenomenon called **global control**. Each control point for the cubic B-spline representation contributes at most to four arcs and so moving a control point will only change the shape of the curve in the vicinity of the point — a phenomenon called **local control**.

6.3 NATURAL CUBIC SPLINES

An altogether different approach to curve drawing is provided by the natural splines. In this we have a set of points (X0,Y0). X1,Y1), ..., (Xm,Ym) in which the Xi are in strictly ascending order, i.e. Xr+1 > Xr for r = 0, 1, ..., m−1. We seek a smooth curve which passes through all the points. The natural cubic spline consists of a set of cubic arcs defined in each interval [Xr, Xr+1] which join smoothly to their neighbours at the data points

(Xr, Yr). For cubic splines this not only means that the slopes of their tangents will be the same, but also their second derivatives at either side of a join will be equal. Only their third derivatives may be discontinuous.

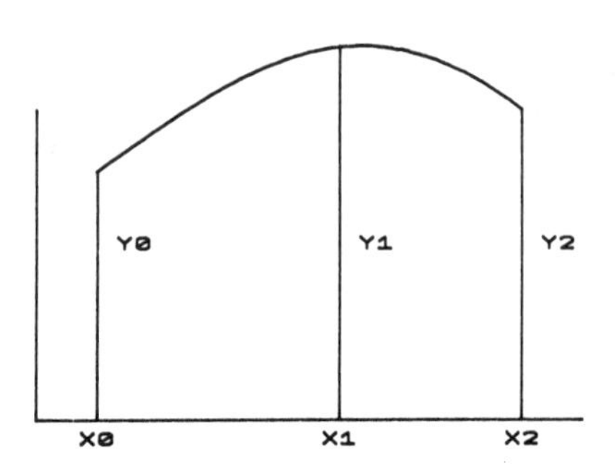

Fig. 6.11.

Since each cubic arc has four degrees of freedom and we only have two conditions to meet for the first arc in [X0, X1] we can impose two further initial conditions at X0, say. Suppose F0 is chosen to be the value of the first derivative of the cubic at x=X0 and S0 is the corresponding value of the second derivative. We can now find T0, the value of third derivative at x=X0 from Taylor's series:

$$Y1 = f(X0+h) = Y0+h.F0+h^2.S0/2+h^3.T0/6 \qquad \text{where } h=X1-X0$$

and we can rearrange this to give an expression for T0 (which is constant in [X0, X1]). Thus we can find the coordinates (X, Y) of any point on the cubic arc joining (X0, Y0) to (X1, Y1) in terms of a parameter t:

$$X = X0+t.h \qquad Y = Y0+t.h.F0+t.h^2.S0/2+t.h^3.T0/6$$

In order to repeat the process for the next cubic arc in [X1, X2] we need to find F1 and S1, the common values of the first and second derivatives at x=X1 and again we can use Taylor's series:

$$F1 = F0+h.S0+h^2.T0/2 \qquad S1=S0+h.T0$$

Now we have all the necessary information: Y1, F1, S1 and Y2 to repeat the process in the inteval [X1, X2] and so on:

```
 10 REM Prog.6.7 - A Natural spline with initial conditions
 49 :
 50 MODE 1 : PROCsetup
 60 OX = 0.1*SW : OY = 0.1*SH
 70 SX = 100 : SY = 100
 99 :
100 n = 6 : m = n-1
```

```
 110 DIM X(m), Y(m), F(m), S(m), T(m)
 120 PROCdefine : F(0) = 2 : S(0) = 0.0006
 130 FOR i = 0 TO m-1
 140  PROCsplinearc(FC)
 150 NEXT i
 490 END
 499 :
1500 DEF PROCdefine
1510  FOR i = 0 TO m
1520   READ X,Y
1530   IF i = 0 THEN X1 = X : Y1 = Y
1540   PROCjoin(X1,Y1, X,Y, 1)
1550   X(i) = X : Y(i) = Y : X1 = X : Y1 = Y
1560  NEXT i
1570  DATA 1,4, 3,6, 4,4.5
1580  DATA 7,8, 9,3, 10,4
1590 ENDPROC
1599 :
1600 DEF PROCsplinearc(PC)
1610  j = i+1 : h = X(j) - X(i) : h2 = h*h/2 : h3 = h2*h/3
1620  Y = Y(i)+h*F(i)+h2*S(i)
1630  T(i) = (Y(j)-Y)/h3
1640  F(j) = F(i)+h*S(i)+h2*T(i)
1650  S(j) = S(i)+h*T(i)
1660  FOR t = 0 TO 1 STEP 1/32
1670   H = t*h : H2 = t*t*h2 : H3 = t*t*t*h3
1680   Y = Y(i)+H*F(i)+H2*S(i)+H3*T(i)
1690   X = X(i) + H
1700   IF t=0 THEN X1 = X : Y1 = Y
1710   PROCjoin(X1,Y1, X,Y, PC)
1720   X1 = X : Y1 = Y
1730  NEXT t
1790 ENDPROC
```

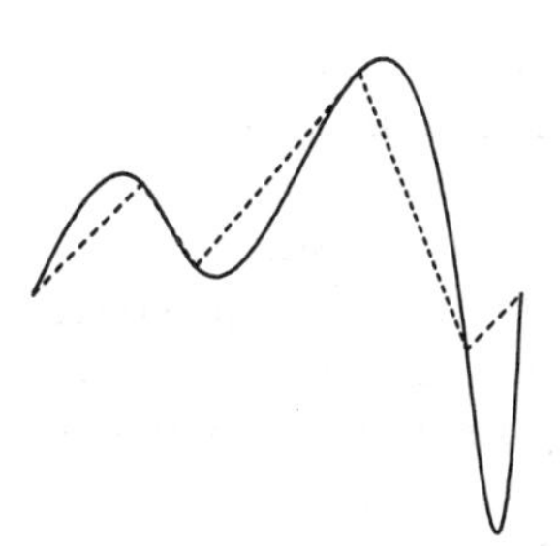

Fig. 6.12.

Experiment with the values of F(0) and S(0) in line 120 to see their effect on the shape of the curve. For example, you could set S(0)=0, print out S(m), and try to adjust F(0) to make S(m)=0.

A more common application of natural cubic splines is where there are **boundary** conditions rather than **initial** conditions to be satisfied. A common problem is that of the 'relaxed' end-conditions in which the value of F(0) is unknown, initially, but it is assumed that S(0)=S(m)=0. In this case we still have enough information to determine each of the cubic arcs but we cannot generate the necessary information for each arc directly as in the last example. Now we have to solve a set of simultaneous equations at each of the internal divisions X1, X2, . . . , Xm−1.

The usual approach is to say that the arcs will be fully defined if we also know the values S1, S2, . . . , Sm−1 of the common second derivatives at all the internal points and to solve a set of simultaneous equations for these. Returning to Taylor's series and considering the two cubic arcs in [Xi−1, Xi] and [Xi, Xi+1] we have:

$$(1): \; Y_{i+1} = Y_i + h.F_i + h^2.S_i/2 + h^3.T_i/6$$
$$(2): \; Y_{i-1} = Y_i - g.F_i + g^2.S_i/2 - g^3.T_{i-1}/6$$
$$(3): \; F_{i+1} = F_i + h.S_i + h^2.T_i/2$$
$$(4): \; F_{i-1} = F_i - g.S_i + g^2.T_{i-1}/2$$
$$(5): \; S_{i+1} = S_i + h.T_i$$
$$(6): \; S_{i-1} = S_i - g.T_{i-1}$$

where $h = X_{i+1} - X_i$ and $g = X_i - X_{i-1}$.

Rearranging (1) and (2) and subtracting gives:

$$(Y_{i+1} - Y_i)/h - (Y_i - Y_{i-1})/g = h.(3.S_i + h.T_i)/6 + g.(3.S_i - g.T_{i-1})/6$$

From (5) and (6) we can find $h.T_i$ and $-g.T_{i-1}$ to give:

$$(7): \; (Y_{i+1} - Y_i)/h - (Y_i - Y_{i-1})/g = h.S_{i-1}/6 + (h+g).S_i/3 + g.S_{i-1}/6$$

Since we know S_0 and S_m, the set of equations (7) for i = 1, 2, . . . , m−1 gives a **tridiagonal** set of simultaneous equations that can be easily solved by a standard numerical algorithm. The following program uses triangular decomposition in PROCmatsolve.

The curve is defined in each **span** once the Y and S values are known for the two endpoints X_{i-1} and X_i. Rather than calculating the corresponding F and T values it is more efficient to express Y in the form:

$$Y = A.(1-t)^3 + B.t^3 + C.(1-t) + D.t$$

and we can find that:

$$A = S_{i-1}.h^2/6 \qquad B = S_i.h^2/6$$
$$C = Y_{i-1} - A \qquad D = Y_i - B$$

where $h = X_i - X_{i-1}$:

```
  10 REM Prog.6.8 - A Natural spline with boundary conditions
  49 :
  50 MODE 1 : PROCsetup
  60 OX = 0.1*SW : OY = 0.1*SH
  70 SX = 100 : SY = 100
  99 :
 100 n = 6 : m = n-1
 110 DIM X(m), Y(m), S(m)
 120 PROCdefine
 130 PROCspline
 140 PROCcurve(FC)
 490 END
 499 :
1500 DEF PROCdefine
1510  FOR i = 0 TO m
1520   READ X,Y
1530   IF i = 0 THEN X1 = X : Y1 = Y
1540   PROCjoin(X1,Y1, X,Y, 1)
1550   X(i) = X : Y(i) = Y : X1 = X : Y1 = Y
1560  NEXT i
1570  DATA 1,4, 3,6, 4,4.5
1580  DATA 7,8, 9,3, 10,4
1590 ENDPROC
1599 :
4000 DEF PROCspline
4010  DIM h(m), B(m), D(m), C(m), Z(m), U(m)
4020  FOR i = 0 TO m-1
4030   h(i) = X(i+1) - X(i)
4050  NEXT i
4060  PROCmatsolve
4090 ENDPROC
4099 :
6000 DEF PROCmatsolve
6010  FOR i = 0 TO m-1
6020   D(i) = (Y(i+1) - Y(i))/h(i)
6030  NEXT i
6040  FOR i = 1 TO m-1
6050   B(i) = (D(i) - D(i-1))*6
6060  NEXT i
6070  C(0) = 1/(2*(X(2) - X(0)))
6080  U(1) = h(1)*C(0)
6090  FOR i = 1 TO m-2
6100   C(i) = 1/(2*(X(i+2) - X(i)) - h(i)*U(i))
6110   U(i+1) = h(i+1)*C(i)
6120  NEXT i
6130  Z(1) = B(1)*C(0)
6140  FOR i = 2 TO m-1
6150   Z(i) = (B(i) - h(i-1)*Z(i-1))*C(i-1)
6160  NEXT i
6170  S(0) = 0 : S(m) = 0
6180  S(m-1) = Z(m-1)
6190  FOR i = m-2 TO 1 STEP -1
6200   S(i) = Z(i) - U(i)*S(i+1)
6210  NEXT i
6900 ENDPROC
6999 :
7000 DEF PROCcurve(PC)
7010  X1 = X(0) : Y1 = Y(0)
7020  dt = 0.05
7030  FOR i = 1 TO m
7040   h = h(i-1) : h = h*h/6
7050   A = S(i-1)*h : B = S(i)*h
7060   C = Y(i-1)-A : D = Y(i)-B
7070   FOR t = dt TO 1.01 STEP dt
7080    u = 1-t
```

```
7090     X = X(i-1) + t*h(i-1)
7100     Y = u*(u*u*A+C)+t*(t*t*B+D)
7110     PROCjoin(X1,Y1, X,Y, PC)
7120     X1 = X : Y1 = Y
7130    NEXT t
7140   NEXT i
7190 ENDPROC
```

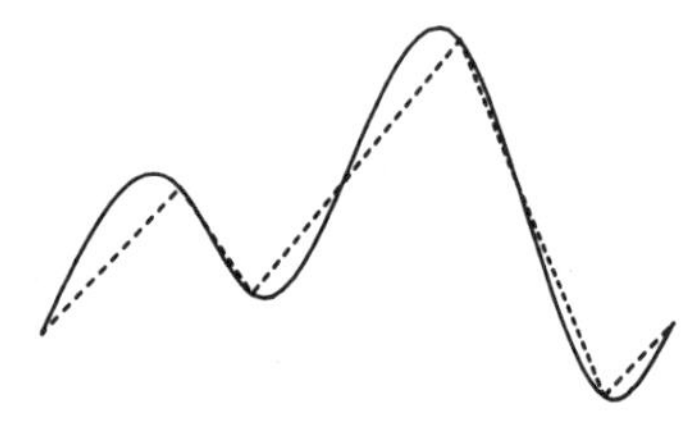

Fig. 6.13.

6.4 NAME DROPPING: HERMITE AND CHEBYCHEV

Now that we have developed a good range of techniques both for numerical calculation and graphical display we can try applying them to some further interesting examples.

6.4.1 The Hermite cubic curve

As with all cubic polynomials this will be determined by four pieces of information. In the case of the Bezier and B-spline cubics these were the coordinates of four points. In the case of the natural cubic spline we used just two points together with values of the second derivative of the curve at each of them. The Hermite cubic similarly uses the coordinates A(Xa,Ya), B(Xb,Yb) of the two end-points together with the tangent vectors **Ta**=(Ia,Ja) and **Tb**=(Ib,Jb) at these points. The curve is given as a parametric 'blend' of these four quantities:

$$\mathbf{P} = (2.t^3-3.t^2+1).\mathbf{A}+(-2.t^3+3.t^2).\mathbf{B}$$
$$+(t^3-2.t^2+t).\mathbf{Ta}+(t^3-t^2).\mathbf{Tb}$$

Thus, like the Bezier and B-spline, but unlike the natural spline, there is no restriction on the Hermite cubic to lie in a plane.

```
 10 REM Prog.6.9 - Hermite cubic curve
 49 :
 50 MODE 1 : PROCsetup
 60 OX = 0.1*SW : OY = 0.1*SH
 70 SX = 100 : SY = 100
 99 :
100 Xa = 1 : Ya = 1
110 Ia = 3 : Ja = 8
120 PROCjoin(Xa,Ya, Xa+Ia,Ya+Ja, 1)
130 Xb = 10 : Yb = 5
140 Ib = -2 : Jb = -3
150 PROCjoin(Xb,Yb, Xb+Ib,Yb+Jb, 1)
160 FOR t = 0 TO 1 STEP 1/16
170  t2 = t*t : t3 = t2*t : f = 2*t3 - 3*t2 + 1
180  g = -2*t3 + 3*t2 : h = t3 - 2*t2 + t : i = t3 - t2
190  X = f*Xa + g*Xb + h*Ia + i*Ib
200  Y = f*Ya + g*Yb + h*Ja + i*Jb
210  IF t=0 THEN X1 = X : Y1 = Y
220  PROCjoin(X1,Y1, X,Y, FC)
230  X1 = X : Y1 = Y
240 NEXT t
490 END
```

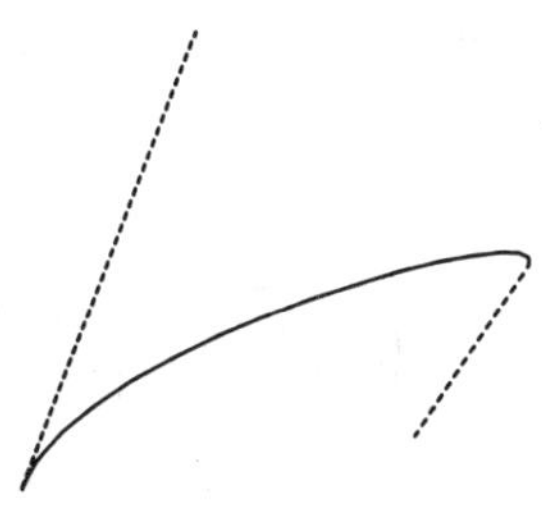

Fig. 6.14.

6.4.2 Chebychev polynomials

Section 14.2 of Oldknow and Smith describes how Chebychev polynomials may be used to refine truncated Taylor (and Maclaurin) series approximations to functions such as COS(x). The n-th Chebychev polynomial is defined as:

$$Tn(x) = COS(n*ACS(x))$$

where ACS is the inverse cosine. The domain of this function is obviously $[-1,1]$ and so is its range. On first sight it is not immediately obvious that Tn(x) is an n-th degree polynomial in x, but furthermore, if n is odd the polynomial only contains odd powers of x and if n is even it only contains even powers. The first few values of Tn(x) are:

$T0(x)=1$
$T1(x)=x$
$T2(x)=2.x^2-1$
$T3(x)=4.x^3-3.x$
$T4(x)=8.x^4-8.x^2+1$

The following program just graphs a few of these:

```
  10 REM Prog.6.10 - Chebychev curves
  49 :
  50 MODE 1 : PROCsetup
  60 OX = 0.5*SW : OY = 0.5*SH
  70 SX = 400 : SY = 400
  99 :
 100 DEF FNcheb(n,X) = COS(n*ACS(X))
 110 :
 120 PROCaxes(1)
 130 FOR n = 1 TO 6
 140  X1 = -1 : Y1 = FNcheb(n,-1)
 150  FOR X = -1 TO 1 STEP 1/32
 160   Y = FNcheb(n,X)
 170   PROCjoin(X1,Y1, X,Y, FC)
 180   X1 = X : Y1 = Y
 190  NEXT X
 200 NEXT n
 490 END
 499 :
2000 DEF PROCaxes(PC)
2010  PROCjoin(-1,0, 1,0, PC)
2020  PROCjoin(0,-1, 0,1, PC)
2090 ENDPROC
```

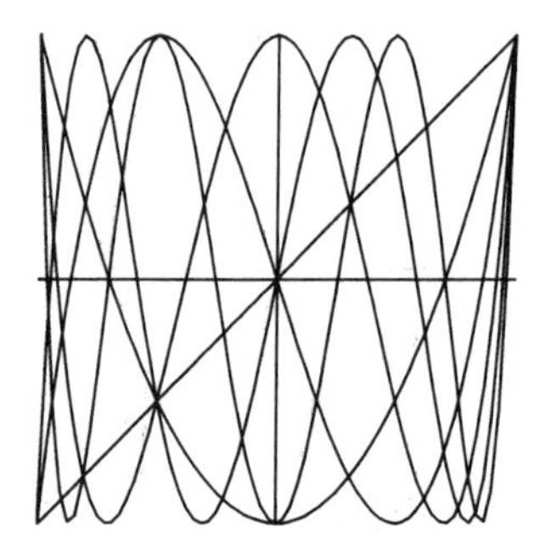

Fig. 6.15.

Try taking a function such as COS(x), graphing it in red, computing successive approximations to it using both Maclaurin and Chebychev series and plotting the results in different colours.

6.5 3D REPRESENTATION — SOME SIMPLE PROJECTIONS

There are several well known projections from three dimensions to two dimensions used in pictorial representation of space; e.g.: orthographic, oblique, isometric and perspective. In each case we need to define a mapping from the 3D coordinates (x,y,z) of a point P in space to the 2D coordinates (Xs,Ys) of its image P′ in the 'picture plane'.

6.5.1 Orthographic projection

In this case one of the three coordinates is just ignored. If the picture plane is taken as the plane x=0 then the mapping simply becomes:

$$Xs=y \qquad Ys=z$$

Consider drawing a vertical triangle OPQ:

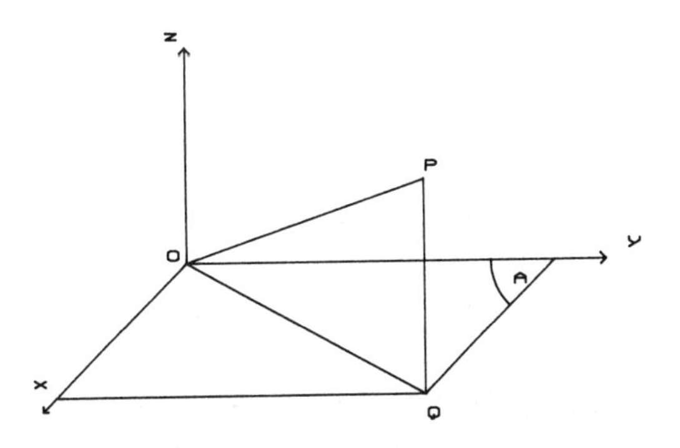

Fig. 6.16.

A simple program to show the projection needs to fix an origin, axes and scale factors to give screen coordinates (Xs,Ys):

```
 10 REM Prog.6.11 - Orthographic projection
 49 :
 50 MODE 1 : PROCsetup
 60 OX = 0.5*SW : OY = 0.5*SH
 70 SX = 50 : SY = 50
 99 :
100 PROCjoin(0,0, 5,0, 1)
110 PROCjoin(0,0, 0,5, 1)
300 x = 5 : y = 7 : z = 4
310 Xs = y : Ys = z
320 PROCjoin(0,0, Xs,Ys, FC)
330 X1 = Xs : Y1 = Ys
340 x = 5 : y = 7 : z = 0
350 Xs = y : Ys = z
360 PROCjoin(X1,Y1, Xs,Ys, FC)
370 PROCjoin(Xs,Ys, 0,0, FC)
490 END
```

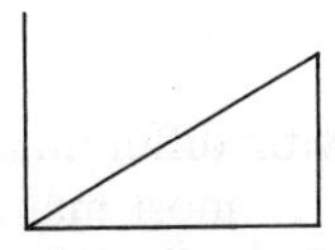

Fig. 6.17.

6.5.2 Oblique projection

In this form the x-axis is conventionally shown drawn at some angle A (usually 45 degrees) going from the middle of the paper towards the bottom left. The y-axis is horizontal and the z-axis vertical. Thus the 2D horizontal coordinate Xs depends upon the 3D y-coordinate and the 2D vertical coordinate Ys depends upon the 3D z-coordinate. To give some effect of 'perspective depth' the distances along the x-axis are usually foreshortened by some factor k. Points that lie in front of the yz-plane (x>0) get swung down and to the left depending upon the 3D x-coordinate. Hence the mapping from 3D to 2D is given by:

$$Xs=y-k*x*COS(A) \qquad Ys=z-k*x*SIN(A)$$

Thus we have the simple relations:

$$Xs=y-m1*x \qquad Ys=z-m2*x$$

where m1=k*COS(A) m2=k*SIN(A)

Only a few changes are needed to Prog. 6.11 to illustrate the vertical triangle in oblique projection. Choices for the foreshortening factor k and the angle of obliqueness A are made so that the multipliers m1 and m2 can be fixed.

```
120 k = 0.7 : A = PI/4
130 m1 = k*COS(A) : m2 = k*SIN(A)
140 PROCjoin(0,0, -5*m1,-5*m2, 1)

310 Xs = y - m1*x : Ys = z - m2*x

350 Xs = y - m1*x : Ys = z - m2*x
```

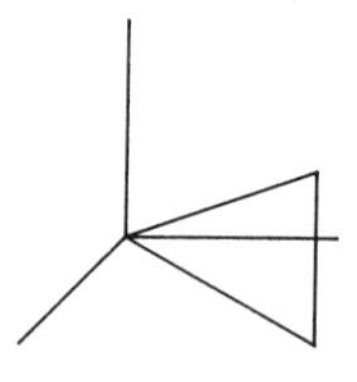

Fig. 6.18.

Try experimenting with different values of k and A. The kind of 3D representation shown in most mathematics books uses m1=m2=0.5. Try different triangles with some of the coordinates negative. Try to represent some 3D solid object like a cuboid or prism.

6.5.3 Isometric projection

In this representation the z-axis remains vertical but the x- and y-axes make equal angles (i.e. 120 degrees) with each other. In this case the screen horizontal component depends upon the difference between the y- and x-coordinates and the vertical component depends upon the z-coordinate and the sum of the x- and y-coordinates. The mapping, of course, uses the sine and cosine of 30 degrees.

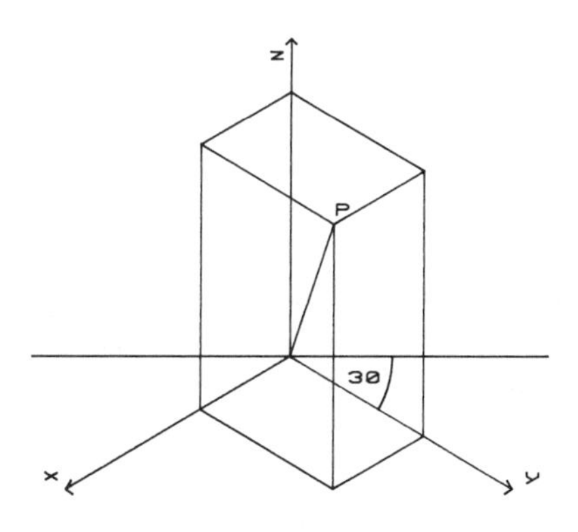

Fig. 6.19.

In this system the projection from (x,y,z) to (Xs,Ys) is:

$$Xs=(y-x)*COS(PI/6) \qquad Ys=z-(x+y)*SIN(PI/6)$$

and the screen coordinates (Xs,Ys) are given by:

$$C=COS(PI/6) \qquad S=SIN(PI/6)$$
$$Xs=(y-x)*C \qquad Ys=z-(x+y)*S$$

Thus the changes to the triangle drawing program are:

```
100 C = COS(PI/6) : S = SIN(PI/6)
105 PROCjoin(0,0, 5*C,-5*S, 1)
120 :
130 :
140 PROCjoin(0,0, -5*C,-5*S, 1)
310 Xs = (y-x)*C : Ys = z - (x+y)*S
350 Xs = (y-x)*C : Ys = z - (x+y)*S
```

6.5.4 Curves in space

As an example of plotting curves in space we can show the trajectory of a missile in oblique projection. We just need to be able to calculate a number

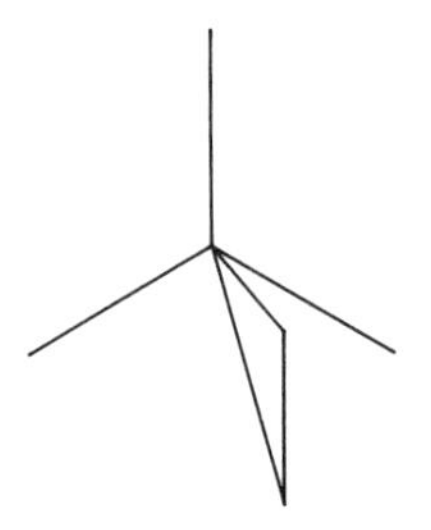

Fig. 6.20.

of points (x, y, z) on the path. These can be projected to screen coordinates (Xs, Ys) and joined by straight lines.

If the projectile starts from the origin we can determine its path from its initial velocity v, its vertical angle of projection av, its horizontal angle of projection ah (measured from the x-axis towards the y-axis) and gravity g.

If we resolve the initial velocity vector into horizontal and vertical components vh and vz we have;

vz=v*SIN(av)   vh=v*COS(av)

and we can resolve vh into components in the x and y directions:

vx=vh*COS(ah)   vy=vh*SIN(ah)

As gravity only affects motion in the z direction we have the equations of motion:

x=vx*t   y=vy*t   z=vz*t−g*t*t/2

Hence the modified program becomes

```
 10 REM Prog.6.12 - A trajectory in 3 dimensions
 49 :
 50 MODE 1 : PROCsetup
 60 OX = 0.3*SW : OY = 0.6*SH
 70 SX = 20 : SY = 20
 99 :
100 PROCjoin(0,0, 10,0, 1)
110 PROCjoin(0,0, 0,10, 1)
120 k = 0.7 : A = PI/4
130 m1 = k*COS(A) : m2 = k*SIN(A)
140 PROCjoin(0,0, -10*m1,-10*m2, 1)
300 X1 = 0 : Y1 = 0
310 v = 20 : g = 10 : ah = PI/3 : av = PI/4
320 t = 0 : dt = 0.1
330 vh = v*COS(av)  : vz = v*SIN(av)
340 vx = vh*COS(ah) : vy = vh*SIN(ah)
350 REPEAT
360  t = t + dt
```

```
370   x = vx*t : y = vy*t
380   z = vz*t - g*t*t/2
390   Xs = y - m1*x : Ys = z - m2*x
400   PROCjoin(X1,Y1, Xs,Ys, FC)
410   X1 = Xs : Y1 = Ys
420 UNTIL z<=0
490 END
```

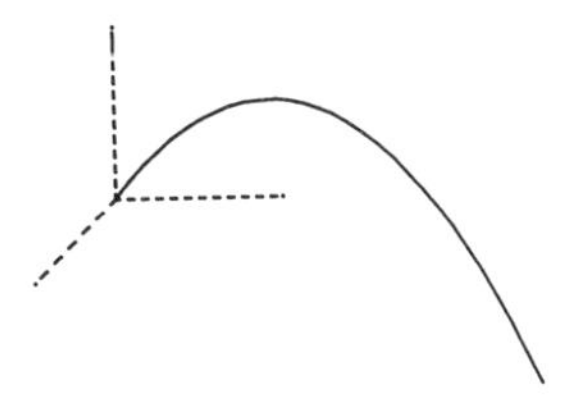

Fig. 6.21.

Try showing the trajectory in other projections. Make the missile bounce on impact with the floor (z=0). Try illustrating other space curves, such as Bezier or B-splines, in 3D.

6.5.5 First-angle projection

In technical drawing a standard representation of a three-dimensional object consists of presenting three orthographic views representing plan, front and side elevations. Section 1.2 of Gasson's *Geometry of Spatial Forms* presents a nice treatment of such representations.

As an exercise in applying the representational techniques developed so far we can try to take a parametric space curve of the form:

$$x=f(t) \qquad y=g(t) \qquad z=h(t)$$

display it in, say, isometric projection and also show the three first-angle orthographic projections.

To do this we need to think about the screen layout. If we draw the isometric picture in the lower left-hand quarter of the screen then we can use the upper left-hand quarter to show the projection on the xz-plane (y=0), the side-elevation. We can use the upper right-hand quarter to show the yz-projection (x=0), the end elevation. And we can use the lower right-hand quarter to show the xy-projection (z=0), the plan view.

Thus we just need a shift of origin between each projection. For the purpose of this display program we shall make a couple of small variations on the standard circular helix:

$$x=a*COS(t) \qquad y=a*SIN(t) \qquad z=c*t$$

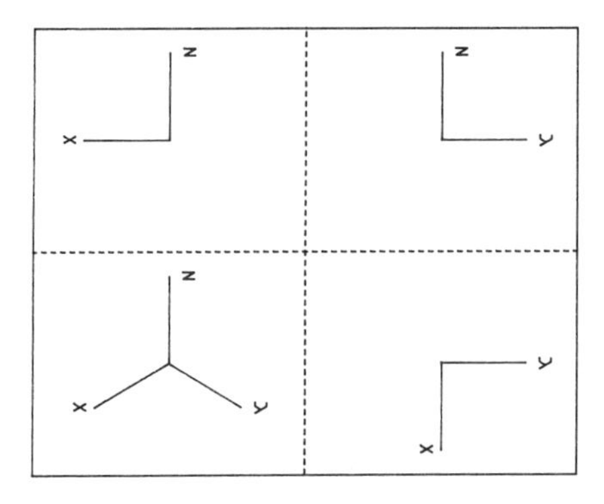

Fig. 6.22.

If we make y=b*SIN(t) then we should have an elliptic cross-section in planes parallel to z=0. We can also make x=a*COS(t)*d(t) where d(t) is some function of t to give an interesting profile in the xz-plane (y=0).

The factors in lines 80, 90 control the horizontal and vertical positions of the orgins for the four screen 'windows'. Try adjusting the factors in lines 70–90 to achieve the best display for your particular micro:

```
  10 REM Prog.6.13 - First angle projection of a space curve
  49 :
  50 MODE 1 : PROCsetup
  60 OX = 0.5*SW : OY = 0.5*SH
  70 SX = 50 : SY = 50
  80 H1 = 0.25 : H2 = 0.75
  90 V1 = 0.25 : V2 = 0.75
  99 :
 100 DEF FNx(t) = a*COS(t)*(1+ABS(t)/5)
 110 DEF FNy(t) = b*SIN(t)
 120 DEF FNz(t) = c*t
 130 a = 1 : b = 1.5 : c = 0.2
 140 tL = -5*PI : tH = 5*PI : tS = (tH-tL)/100
 150 L = 3
 160 PROCxyz
 170 PROCxz
 180 PROCyz
 190 PROCxy
 200 END
 499 :
1500 DEF PROCxyz
1510   OX = H1*SW : OY = V1*SH
1520   S = SIN(PI/6) : C = COS(PI/6)
1530   PROCjoin(0,0, 0,L, 1)
1540   PROCjoin(0,0, L*C,-L*S, 1)
1550   PROCjoin(0,0, -L*C,-L*S, 1)
1560   FOR t = tL TO tH STEP tS
1570     x = FNx(t) : y = FNy(t) : z = FNz(t)
1580     Xs = C*(y-x) : Ys = z - S*(x+y)
1590     IF t=tL THEN X1 = Xs : Y1 = Ys
1600     PROCjoin(X1,Y1, Xs,Ys, FC)
1610     X1 = Xs : Y1 = Ys
1620   NEXT t
1690 ENDPROC
```

```
1699 :
1700 DEF PROCxz
1710   OX = H1*SW : OY = V2*SH
1720   PROCjoin(0,0, -L,0, 1)
1725   PROCjoin(0,0, 0,L, 1)
1730  FOR t = tL TO tH STEP tS
1740   x = FNx(t) : y = FNy(t) : z = FNz(t)
1750   Xs = -x : Ys = z
1760   IF t=tL THEN X1 = Xs : Y1 = Ys
1770   PROCjoin(X1,Y1, Xs,Ys, FC)
1775   X1 = Xs : Y1 = Ys
1780  NEXT t
1790 ENDPROC
1799 :
1800 DEF PROCyz
1810  OX = H2*SW : OY = V2*SH
1820  PROCjoin(0,0, L,0, 1)
1825  PROCjoin(0,0, 0,L, 1)
1830  FOR t = tL TO tH STEP tS
1840   x = FNx(t) : y = FNy(t) : z = FNz(t)
1850   Xs = y : Ys = z
1860   IF t=tL THEN X1 = Xs : Y1 = Ys
1870   PROCjoin(X1,Y1, Xs,Ys, FC)
1875   X1 = Xs : Y1 = Ys
1880  NEXT t
1890 ENDPROC
1899 :
1900 DEF PROCxy
1910  OX = H2*SW : OY = V1*SH
1920  PROCjoin(0,0, L,0, 1)
1925  PROCjoin(0,0, 0,-L, 1)
1930  FOR t = tL TO tH STEP tS
1940   x = FNx(t) : y = FNy(t) : z = FNz(t)
1950   Xs = y : Ys = -x
1960   IF t=tL THEN X1 = Xs : Y1 = Ys
1970   PROCjoin(X1,Y1, Xs,Ys, FC)
1975   X1 = Xs : Y1 = Ys
1980  NEXT t
1990 ENDPROC
```

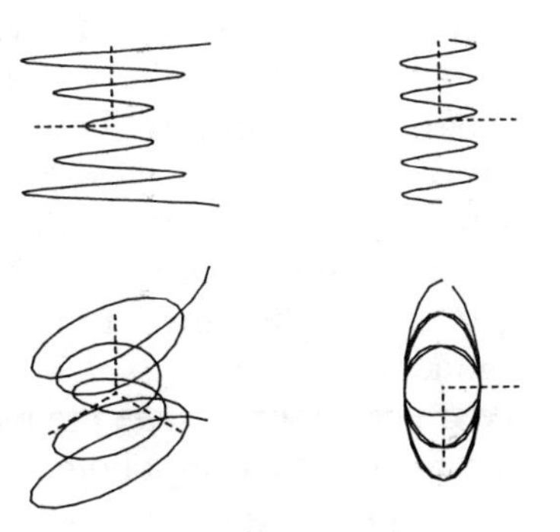

Fig. 6.23.

Try making first-angle projections of other space curves. Find out what differences are needed to display 'third-angle' projections.

7

Surfaces

Surfaces are all about us: car bodies, tea-pots, leaves, tables, sleeves, skin. With the advent of photography, both still and moving, we have become used to a very sophisticated representation of surfaces as slabs of colour and texture whose humps and lumps and curves and things are revealed to our trained eyes by differing qualities of reflected light.

In painting and drawing one of the classic exercises has been to show "still life" which involves trying to show the form and texture of surfaces such as on fruit, crockery, glass and furniture. Those who have tried this know how demanding it is and how sensitive and critical is the eye. Although there are computer systems capable of modelling this kind of surface representation they are enormously sophisticated, powerful and expensive and so well beyond the reach of the microcomputer user, at least for a year or two yet.

The early computer graphic systems were largely developed by rich companies concerned with the making of cars, planes and ships and thus surface representation was a priority. The approach adopted was analogous to the approach we have used to show curves. In this the arc of a curve between two points corresponding to values of x and x+dx is approximated by a straight line, and thus the "curve" is really a polygonal approximation. For example, for the resolution of a micro's display screen, the typical "circle" is a 40-gon.

For a surface we can approximate the piece of surface bounded by the lines joining the **four** points corresponding to x,y values of (x,y), (x+dx,y), (x+ds,y+dy), x,y+dy) by a (possibly twisted) space quadrilateral called a **patch** — named because the final effect is that of a patchwork quilt. Thus the surface is approximated by a kind of polyhedron which is probably easiest thought of as a surface formed by moulding chicken-wire.

In fact a common way of producing model surfaces, e.g. for model railways or dioramas, is to mould chicken-wire and to cover it with wet

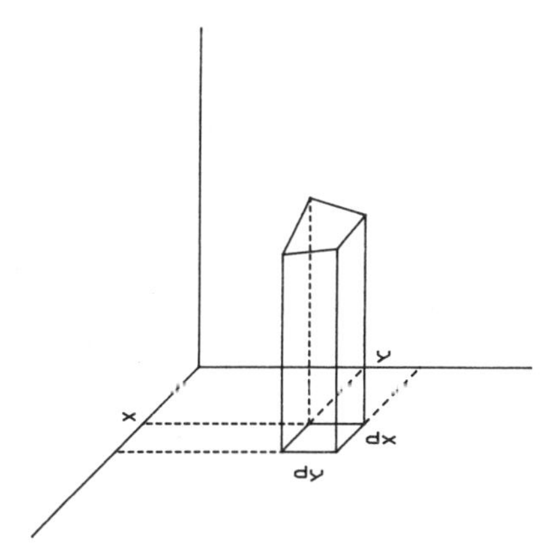

Fig. 7.1 — Diagram of a patch.

papier mâché. The computer produces the chicken-wire and the eye must try to produce the papier mâché!

Figure 7.2 shows the output from a microcomputer representation of a

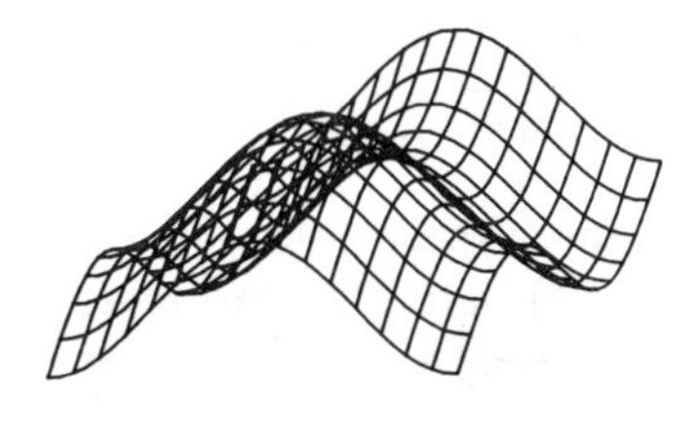

Fig. 7.2 — Output of computer drawn surface.

surface. You may have seen such pictures used on commerical television advertising where it seems to have become quite a trendy way of hinting a company is up-to-the-minute in its technical developments.

The mathematical form of a surface can be expressed, as in the case for curves, in a number of different ways. For example the 3D cartesian coordinates (x,y,z) of a point on a surface might all be functions of just **two** parameters u and v, say, or z might be a function of x and y, or there might be some implicit relation connecting x,y and z. There are other common coordinate systems, too, such as cylindrical polars and spherical polars.

In representing the surface we will have to choose some convention for the positions of the axes and to pick some suitable projection from 3D space coordinates to the 2D representation on the screen.

7.1 A SIMPLE SURFACE: z=f(x,y)

If z can be written as an explicit function of x and y then we can adopt a particularly simple method for representing the surface. Suppose we fix a value for x and let y vary. The corresponding points lie on a curve. if we now change x a little and repeat the process we obtain another curve which lies in a plane parallel to the previous one. If we continue this for a range of values of x we obtain a set of curves corresponding to slices through the surface parallel to the yz-plane. This is what a meat-slicer does to a joint of cold meat.

If we performed the whole process again with the roles of x and y reversed the "slices" would be in planes at right angles to the previous ones.

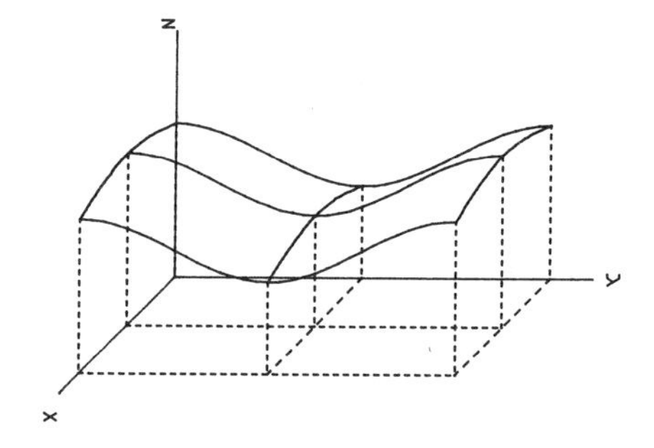

Fig. 7.3 — Diagram of patches.

In order to show these slices on the screen we need to map the 3D coordinates (x, y, z) onto 2D coordinates by some projection. One of the simplest but most effective projections is a 45 degree oblique projection to (Xs,Ys) given by:

$$XS=y-x/2 \qquad Ys=z-x/2$$

Of course we will need, as usual, to shift the origin and scale the result to obtain a satisfactory image on the screen.

In the following program the function of two variables is defined in line 100, the values to fix the ranges of x and y appear in lines 110 and 120 and the display parameters are set in lines 50–70. The double loop in lines 200–270 draws the curves for fixed values of x and the similar loop in lines 300–370 draws the curves for fixed values of y. The inner loop that draws each curve has been deliberately given a smaller step than the outer loop to make the curves appear smoother:

```
10 REM Prog.7.1 - Simple surface representation
49 :
50 MODE 1 : PROCsetup
60 OX = 0.5*SW : OY = 0.5*SH
```

```
 70 SX = 100 : SY = 100
 99 :
100 DEF FNz(x,y) = SIN(x) + COS(y)
110 xl = -PI : xh = PI : dx = (xh-xl)/32
120 yl = -PI : yh = PI : dy = (yh-yl)/32
200 FOR x = xl TO xh STEP 2*dx
210   X1 = yl - x/2 : Y1 = FNz(x,yl) - x/2
220   FOR y = yl TO yh STEP dy
230     Xs = y - x/2 : Ys = FNz(x,y) - x/2
240     PROCjoin(X1,Y1, Xs,Ys, FC)
250     X1 = Xs : Y1 = Ys
260   NEXT y
270 NEXT x
290 :
300 FOR y = yl TO yh STEP 2*dy
310  X1 = y - xl/2 : Y1 = FNz(xl,y) - xl/2
320   FOR x = xl TO xh STEP dx
330     Xs = y - x/2 : Ys = FNz(x,y) - x/2
340     PROCjoin(X1,Y1, Xs,Ys,FC)
350     X1 = Xs : Y1 = Ys
360   NEXT x
370 NEXT y
490 END
499 :
500 DEF PROCsetup
510  SW = 1280 : SH = 1024
520  NC = 3 : FC = 3
590 ENDPROC
599 :
600 DEF PROCdot(X,Y,PC)
610  GCOL 0,PC
620  PLOT 69, OX + X*SX, OY + Y*SY
690 ENDPROC
699 :
700 DEF PROCjoin(X1,Y1, X,Y, PC)
710  GCOL 0,PC
720  MOVE OX + X1*SX, OY + Y1*SY
730  DRAW OX + X*SX, OY + Y*SY
790 ENDPROC
799 :
```

The output from this program was shown in Fig. 7.2.

Figure 7.4 shows some surfaces produced from this program. The first two are from z=SIN(x+y) and z=SIN(x)*COS(y), see if you can guess the form of the others.

As with the case of curves you can be quite adventurous in the way you define your functions, using INT you can have step-functions and using the logical relations <, =, > together with AND, OR and NOT you can have piecewise definitions. Figure 7.5 shows some of these more exotic creations.

Try exploring some other forms of your own.

7.2 PARAMETRIC SURFACES

Program 7.1 displays explicit cartesian surfaces of the form z=f(x,y). In Chapter 3 we saw that for many display purposes the **parametric** representation of a curve was more appropriate. In this section, then, we shall try to extend the idea of parametric representation to a surface. As we are working

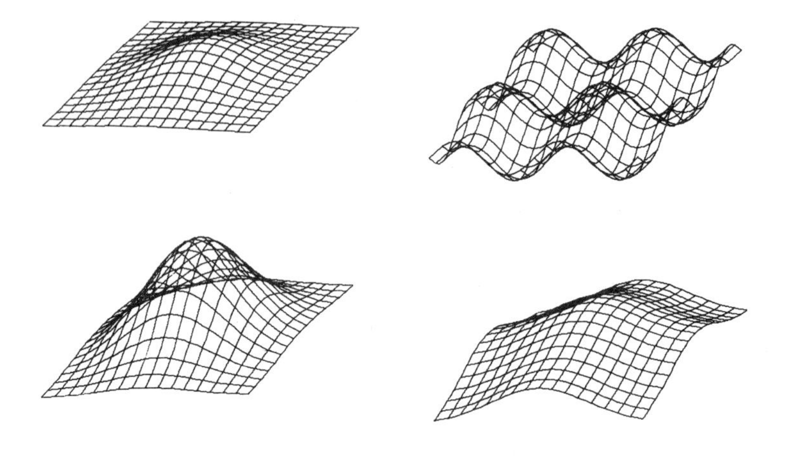

Fig. 7.4 — Output from program 7.1.

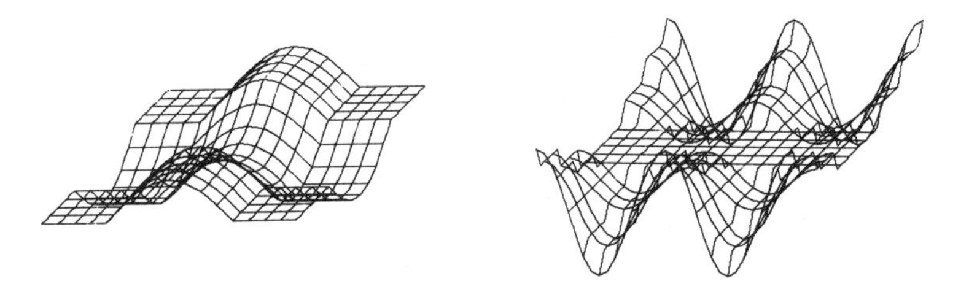

Fig. 7.5 — Some piecewise defined surfaces.

in three dimensions we need the x-, y- and z-coordinates of each point of the surface to be functions of one or more parameters. However, if only one parameter is involved then the points have only one degree of freedom, and so will lie on a curve in space, as in the previous chapter. A surface is a two-dimensional object and so we need exactly two independent parameters, which are conventionally called u and v.

In order to display the surface:

$$x=f(u,v) \qquad y=g(u,v) \qquad z=h(u,v)$$

we must choose some projection from 3D to 2D. Then, holding the parameter u fixed at some suitable value, we can draw the space curve formed as v varies. Then, by continually changing u a little and repeating the process of drawing space curves for changing v, we obtain a family of curves corresponding to different values of u. Similarly by reversing the roles of u and v and repeating the process we obtain another family of curves corresponding to different values of v. The intersections of neighbouring pairs of each family are the quadrilateral **patches** defined by (u,v), (u+du,v), (u+du, v+dv), (u,v+dv) corresponding to the cartesian patches of the last section.

The program to display such a surface is very similar to Program 7.1 except that the nested loops will have u and v as their control variables instead of x and y:

```
 10 REM Prog.7.2    Parametric surface representation
 49 :
 50 MODE 1 : PROCsetup
 60 OX = 0.5*SW : OY = 0.5*SH
 70 SX = 200 : SY = 200
 99 :
100 DEF FNx(u,v) = SIN(u)
110 DEF FNy(u,v) = SIN(v)
120 DEF FNz(u,v) = SIN(u+v)
130 ul = -PI : uh = PI : du = (uh-ul)/32
140 vl = -PI : vh = PI : dv = (vh-vl)/32
190 :
200 FOR u = ul TO uh STEP 2*du
205  x = FNx(u,vl) : y = FNy(u,vl) : z = FNz(u,vl)
210  X1 = y - x/2 : Y1 = z - x/2
220  FOR v = vl TO vh STEP dv
225   x = FNx(u,v) : y = FNy(u,v) : z = FNz(u,v)
230   Xs = y - x/2 : Ys = z - x/2
240   PROCjoin(X1,Y1, Xs,Ys, FC)
250   X1 = Xs : Y1 = Ys
260  NEXT v
270 NEXT u
290 :
300 FOR v = vl TO vh STEP 2*dv
305  x = FNx(ul,v) : y = FNy(ul,v) : z = FNz(ul,v)
310  X1 = y - x/2 : Y1 = z - x/2
320  FOR u = ul TO uh STEP du
325   x = FNx(u,v) : y = FNy(u,v) : z = FNz(u,v)
330   Xs = y - x/2 : Ys = z - x/2
340   PROCjoin(X1,Y1, Xs,Ys,FC)
350   X1 = Xs : Y1 = Ys
360  NEXT u
370 NEXT v
490 END
```

The output from this program is shown in Fig. 7.6.

Again, by changing the definitions in lines 100–120, many different surfaces can be explored. To change the range of the parameters then lines 130, 140 need to be altered. To change the projection we just need different expressions for screen coordinates in terms of x, y and z in lines 210, 230, 310 and 330.

To find the parametric equations of a sphere, radius R, we can arrange for curves given by u=constant to correspond with lines of latitude and those with v=constant to be the lines of longitude. Thus when u ranges from −PI/2 through 0 to PI/2 we want the curves to be circles with radii that grow from 0 to R and then decrease again to 0, and the coordinates of their centres to change from (0,0,−R) through (0,0,0) to (0,0,R). A form which has this

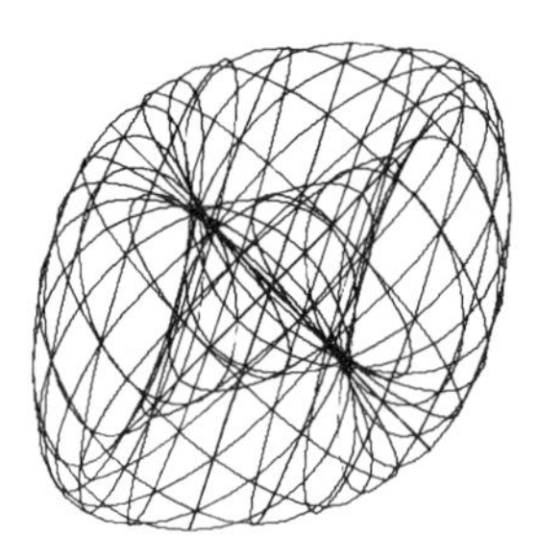

Fig. 7.6 — Output from program 7.2.

property is:

X=R*COS(U)*COS(V)   Y=R*COS(U)*SIN(V)   Z=R*SIN(U)

The changes needed to Program 7.2 are quite small:

```
130 ul = -PI : uh = PI : du = (uh-ul)/32
140 vl = -PI : vh = PI : dv = (vh-vl)/32
150 R = 1
100 DEF FNx(u,v) = R*COS(u)*COS(v)
110 DEF FNy(u,v) = R*COS(u)*SIN(v)
120 DEF FNz(u,v) = R*SIN(u)
```

In fact, the use of R=1 is really redundant and we could just leave the multipliers SX, SY to sort the whole process out. However, this same program structure will allow us, for example, to produce rugby balls (ellipsoids of revolution), ring doughnuts (toruses) and other exotica.

The ellipsoid is given by:

X=A*COS(U)*COS(V)   Y=A*COS(U)*SIN(V)   Z=B*SIN(U)

and the torus by:

X=(A+B*COS(U))*COS(V)   Y=(A+B*COS(U))*SIN(V)   Z=B*SIN(U)

The output corresponding to these three surfaces of revolution is shown in Fig. 7.7. For good measure a "pretzel" surface is also shown — see if you can determine its parametric form.

7.3 CURVES IN A SURFACE

As we have seen, a space curve can be defined by making the x-, y- and z-coordinates of points each be functions of a single parameter. In the last section we drew curves on surfaces by keeping one or other of the u and v surface parameters constant. Other curves can be described by making u and

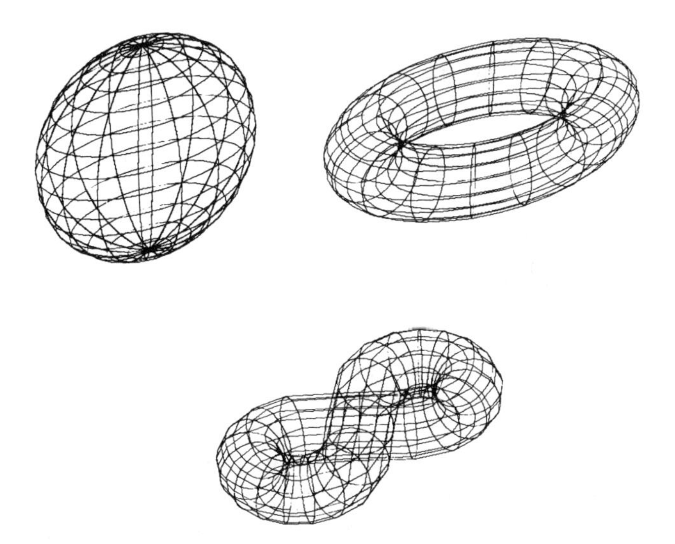

Fig. 7.7 — Output from program 7.2.

v conform to some relation. In particular, if v can be expressed as a function of u, then the corresponding curve can be plotted by a single FOR-NEXT loop with u as the control variable. In the following additions to Program 7.2 v is defined as a function of u, a new range for u is defined and the display colour is changed before the curve is plotted:

```
400 DEF FNv(u) = u
405 ul = 0 : uh = PI/2 : du = (uh-ul)/32
410 FOR u = ul TO uh STEP 2*du
415  v = FNv(u)
420  x = FNx(u,v) : y = FNy(u,v) : z = FNz(u,v)
425  Xs = y - x/2 : Ys = z - x/2
430  IF u = ul THEN X1 = Xs : Y1 = Ys
435  PROCjoin(X1,Y1, Xs,Ys, 2)
440  X1 = Xs : Y1 = Ys
445 NEXT u
```

Fig. 7.8 shows the result of plotting this curve on a torus. Try finding other

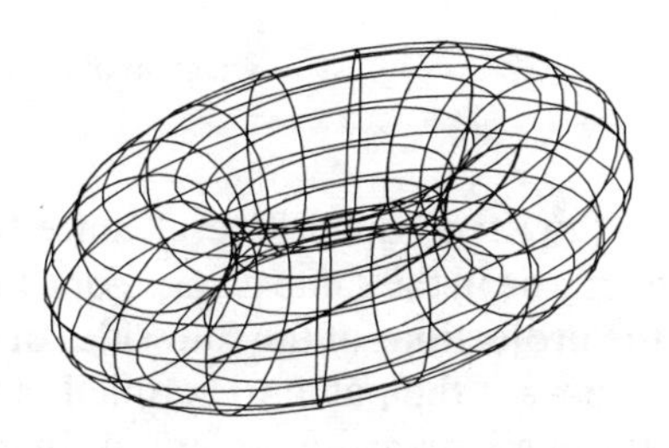

Fig. 7.8 — A curve on a torus.

interesting curves on surfaces. See, for example, if you can draw a helix on a cylinder — like the thread on a screw.

7.4 PATCHES, SURFACE AREA AND VOLUME

In these next three section we shall build up a program in a modular fashion that is capable of displaying a surface using one of a number of graphical techniques, in one of a number of projections while also calculating some quantities. The version chosen will be based on the cartesian representation of a surface: z=f(x,y) and the changes needed to make this handle parametric surfaces should by now be obvious.

The starting point is the notion of a surface **patch.** In the representations of section 7.1 and 7.2 the surface is eventually formed by sets of lines which divide the display up into little quadrilateral regions, like the holes in a net. Such little regions are called **patches** and the whole surface can be considered as made up from many such patches "stitched" together like a patchwork quilt.

If dx and dy are the step lengths along the axes, then a typical patch will be the piece of the surface that lies directly above or below the rectangle in the x,y-plane whose corners are (x,y,0), (x,y+dy,0), (x+dx,y+dy,0) and (x+dx,y,0) as shown in Fig. 7.9. For small dx and dy the actual patch can be

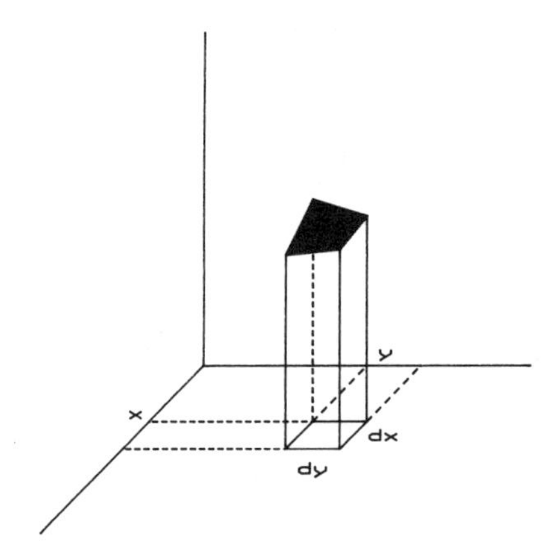

Fig. 7.9 — Diagram of a patch.

considered as a "twisted" quadrilateral, since the four corners of the patch will not necessarily be in the same plane. For a line display we just need to be able to draw the projections of the four sides of the twisted quadrilateral on the screen. If we want the patch to be filled in with colour then we need a procedure that can fill in any quadrilateral. In general this is quite slow and awkward but some computers, such as the BBC micro, provide a standard procedure for filling in any triangle, and others to "flood" fill any polygon. If we approximate the patch by two triangles with a common side then we can easily fill it with colour on a micro that has a triangle-filling procedure. If we take care to fill in the surface starting from the back and working to the front of the display we can arrange to over-paint any hidden parts of the surface.

Whether or not we can display the patch as two triangles we can calculate the areas of each and thus have an approximation to the surface area of the patch. By accumulating these areas we have an approximation to the total surface area. Similarly if we approximate the volume between the patch and its projection on the xy-plane as a cuboid we know its length and breadth to be dx and dy and we can take its height as the z-coordinate of the middle point of the path: z=f(x+dx/2,y+dy/2). By accumulating such elements of volume we can obtain a simple approximation to the volume contained between the surface and the xy-plane.

The program will make use of the following procedures:

PROCproject which projects the space point (x,y,z) to screen coordinates (Xs,Ys) by the chosen projection.
PROCpatch which displays the quadrilateral patch whose back left corner is at (x,y,z) in the chosen representation.
PROCcalculate which accumulates the approximations to the computed quantities such as area and volume.
PROCconstants which calculates the constants used in the program.

In order to avoid duplicating the calculation of some values it is convenient to store screen coordinates in arrays x and y.

In the following program a general patch will be represented on the screen by a quadrilateral whose screen coordinates are as follows:

Back left corner: x(i−1), y(i−1)
Back right corner: x(i), y(i)
Front left corner: X0,Y0
Front right corner: Xs, Ys

```
 10 REM Prog.7.3 - Surface patches
 49 :
 50 MODE 1 : PROCsetup
 60 REM OX,OY,SX,SY are defined in PROCconstants
 99 :
100 DEF FNz(x,y) = SIN(x) + COS(y)
110 xl = -PI : xh = PI
120 yl = -PI : yh = PI
130 zl = -2  : zh = 2
140 PROCconstants
145 :
150 FOR x = xl TO xh STEP dx
160   FOR i = 0 TO ny
170     y = yl + i*dy : z = FNz(x,y)
180     PROCproject
190     IF x>xl AND i>0 THEN PROCpatch
200     IF i>0 THEN x(i-1) = X0 : y(i-1) = Y0
210     X0 = Xs : Y0 = Ys
```

```
  220  NEXT i
  230  x(i-1) = Xs : y(i-1) = Ys
  240 NEXT x
  250 PRINT "Volume = ";vol
  490 END
  499 :
  999 REM******** constants **********
 1000 DEF PROCconstants
 1010  nx = 32 : ny = 32 : dx = (xh-xl)/nx : dy = (yh-yl)/ny
 1020  DIM x(nx), y(ny)
 1030  vol = 0
 1040  xmin = yl-xh/2 : xmax = yh-xl/2
 1050  ymin = zl-xh/2 : ymax = zh-xl/2
 1060  SX = SW/(xmax - xmin)
 1070  SY = SH/(ymax - ymin)
 1080  OX = -xmin*SX : OY = -ymin*SY
 1090 ENDPROC
 1095 :
 1099 REM******** project ************
 1100 DEF PROCproject
 1110  Xs = y - x/2
 1120  Ys = z - x/2
 1190 ENDPROC
 1195 :
 1199 REM******** patch **************
 1200 DEF PROCpatch
 1210  PROCcalculate
 1220  PROCfill (x(i),y(i), x(i-1),y(i-1), Xs,Ys, 1)
 1230  PROCfill(x(i-1),y(i-1), Xs,Ys, X0,Y0, 1)
 1240  PROCjoin(x(i),y(i), x(i-1),y(i-1), FC)
 1250  PROCjoin(x(i-1),y(i-1), X0,Y0, FC)
 1260  PROCjoin(X0,Y0, Xs,Ys, FC)
 1270  PROCjoin(Xs,Ys, x(i),y(i), FC)
 1290 ENDPROC
 1295 :
 1299 REM******** calculate **********
 1300 DEF PROCcalculate
 1310  dv = dx*dy*FNz(x-dx/2,y-dy/2)
 1320  vol = vol + dv
 1390 ENDPROC
 1395 :
 9999 REM******** fill a triangle *****
10000 DEF PROCfill(X2,Y2, X1,Y1, X,Y, PC)
10010  GCOL 0,PC
10020  MOVE OX + X2*SX, OY + Y2*SY
10030  MOVE OX + X1*SX, OY + Y1*SY
10040  PLOT 85, OX + X*SX, OY + Y*SY
10090 ENDPROC
10099 :
```

The constants nx and ny fix how many divisions are to be taken along each axis — thus there are nx*ny patches. In this particular version of the program the chosen projection is a 45 degree oblique projection and the screen constants SX, SY, OX and OY are chosen so as to position the display to occupy as much of the screen as possible. In this version each patch is first filled with a colour (red) in such a way as to obliterate any previous drawing "underneath" it, and then its outline is drawn in a different colour (white). Thus the finished effect has removed any hidden detail. The calculations performed in this version of the program only collect information about the volumes. Later we shall see how to extend this to surface area.

This program will convert in its entirety for other micros if a routine can be defined for PROCfill, otherwise lines 1220 and 1230 will need to be omitted.

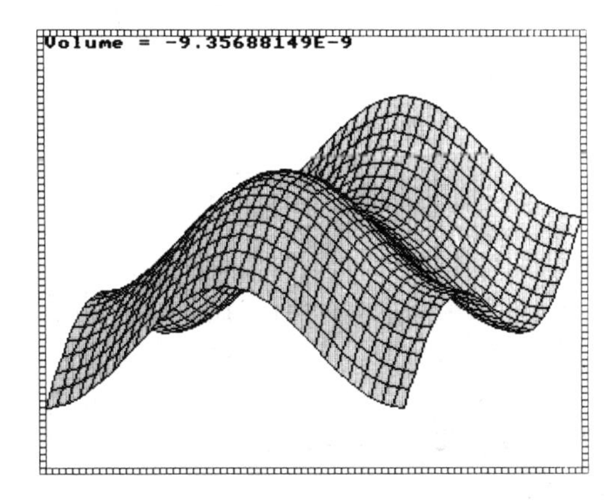

Fig. 7.10 — Output from program 7.3.

The value for the volume contained between this surface and the xy-plane is computed to be:

Volume=−9.35688149E-9

By symmetry (or by double integration) we can see that the exact result is zero. The fact that we have obtained a numerical approximation astonishingly close to the analytic result must be regarded as a bit of a fluke!

COLOURING A SURFACE

Our representation of the surface z=f(x,y) is made up from **patches** formed by the four points P:(x,y,zp), Q:(x+dx,y,zq), R:(x,y+dy,zr) and S:(x+dx,y+dy,zs). The normal vector to the patch at P:(x,y,z) can be found from the vector product of **PQ** and **PR. PQ** has components: (dx,0,zq−zp) and **PR** has components: (0,dy,zr−zp). Thus the 'outward' normal at P to the patch has components: (−dy.(zq−zp), −dx.(zr−zp), dx.dy). The dot product of this with a viewing vector can be used to find whether the top of the patch is visible or not.

To find the viewing vector for a particular projection we just need to establish the general form of the coordinates of points (x,y,z) that project to the origin of the 2D representation (0,0) — this is the **kernel** of the projection. For an oblique projection we have the mapping:

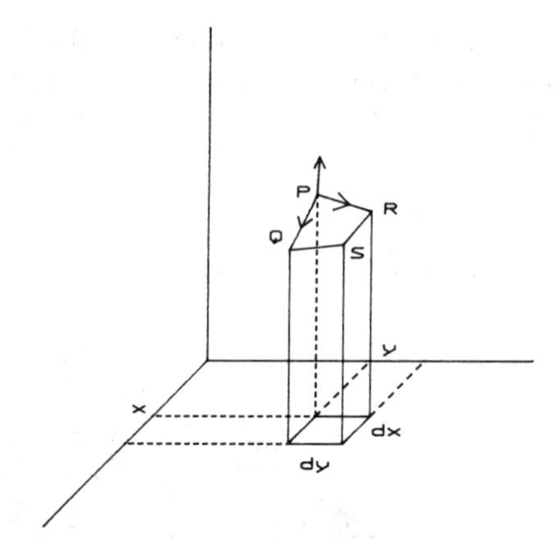

Fig. 7.11 — Diagram of patch.

Xs=y−x/2   Ys=z−x/2

Thus any point whose coordinates are in the ratio:

x:y:z=2:1:1

will map onto the screen origin. Thus the vector with components (2,1,1) will serve as a viewing vector. Taking the dot product of this with the normal to the surface patch gives a value which discriminates between the top and bottom of our view of the patch. If we define the variable 'dot' by:

dot=−2∗dy∗(zq−zp)−dx∗(zr−zp)+dx∗dy

then we are looking at one side of the surface at P if dot>0 and the other if dot<0. If dot=0 then we are viewing the surface 'edge on' at P.

The changes to the program are slight. Instead of using a constant pen-colour in lines 1220, 1230 we can determine the colour to be used for the patch, depending on whether we are viewing the top or the bottom of the surface.

```
1215   PROCcolour
1220   PROCfill (x(i),y(i), x(i-1),y(i-1), Xs,Ys, PC)
1230   PROCfill(x(i-1),y(i-1), Xs,Ys, X0,Y0, PC)
1395 :
1399 REM******** colour ************
1400 DEF PROCcolour
1410   zq = FNz(x+dx,y) : zr = FNz(x,y+dy)
1420   dot = -2*dy*(zq-z) - dx*(zr-z) + dx*dy
1430   IF dot>0 THEN PC = 1 ELSE PC = 2
1490 ENDPROC
1495 :
```

In fact, for our chosen test surface, this effect is only really impressive in its dynamic form — seeing the surface built up in front of your eyes. The finished form of the surface is virtually all displayed in just one colour so you are urged to try the effect on your own computer.

7.6 CALCULATING SURFACE AREA

Knowing the space coordinates of the four corners P, Q, R, S of each surface patch we should be able to compute approximations for the area of each patch. By accumulating these we will be able to obtain an approximation to the surface area of the complete surface. The technique used here is one called **vector area.**

We can approximate the (twisted) quadrilateral patch PQRS by two planar triangles PRS and PRQ. In the triangle PRS let the side **PR** be represented by the vector **a** and the side **PS** by vector **b**. Now one form for the area of a triangle is:

'half the product of two sides and the sine of the included angle'.

If we denote the angle RPS by C then the area of PRS is:

a.b.sin(C)/2

Now the vector product **a**×**b** of vectors **a** and **b** is a vector whose magnitude is a.b.sin(C). Thus if we know the components of the vectors **a** and **b** we can form their vector (or cross) product **t** and then find the magnitude of **t**.

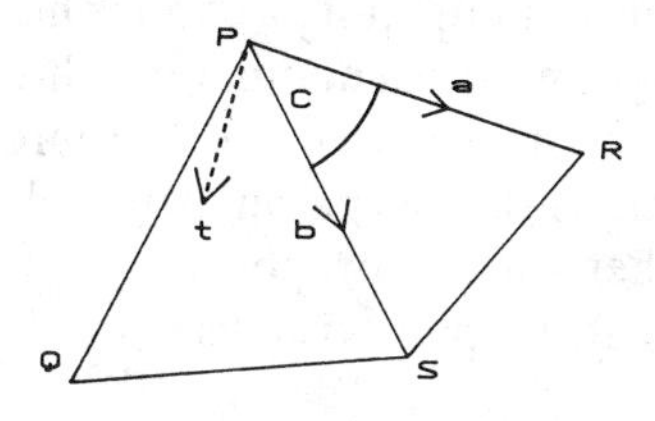

Fig. 7.12 — Vector quadrilateral.

The following procedure calculates an approximation to the area of the patch PQRS using the values zq and zr already calculated in PROCcolour:

```
1499 REM********* area ***************
1500 DEF PROCarea
1510  zs = FNz(x+dx,y+dy)
1520  A(1) = 0  : A(2) = dy : A(3) = zr-z
1530  B(1) = dx : B(2) = dy : B(3) = zs-z
1540  C(1) = dx : C(2) = 0  : C(3) = zq-z
1550  REM Find vector area of first triangle PRS
```

```
1560  tx = A(2)*B(3) - A(3)*B(2)
1570  ty = A(3)*B(1) - A(1)*B(3)
1580  tz = A(1)*B(2) - A(2)*B(1)
1590  t1 = 0.5*SQR(tx*tx + ty*ty + tz*tz)
1600  REM Find vector area of second tr6angle PSQ
1610  tx = B(2)*C(3) - B(3)*C(2)
1620  ty = B(3)*C(1) - B(1)*C(3)
1630  tz = B(1)*C(2) - B(2)*C(1)
1640  t2 = 0.5*SQR(tx*tx + ty*ty + tz*tz)
1650  area = area + t1 + t2
1690 ENDPROC
1695 :
```

To link this with the main program we need only to dimension the vectors A, B and C, to initialise the value of 'area' and to call the procedure within PROCpatch:

```
1035  DIM A(3),B(3),C(3) : area = 0

 260 PRINT "Area = ";area

1280  PROCarea
```

The value obtained for the surface under consideration:

$$z=SIN(x)+COS(y) \qquad -PI \leqslant x \leqslant PI \qquad -PI \leqslant y \leqslant PI$$

is: 55.3306764.

The reader is left to struggle with the analysis to confirm this result. Incidentally the principle of vector area of triangles can be easily applied to any polygon by splitting it into successive triangles with common sides in the way we have just done for a quadrilateral.

We have now built up quite a comprehensive surface representation and calculation program from a number of self-contained procedures. There are many possible extensions: to include different projections, to handle different representations such as parametric, cylindrical and spherical polars and to calculate other surface properties. Certainly this kind of program is an invaluable tool in helping visualise problems in analytic and differential geometry.

8

Solids

8.1 WIREFRAME REPRESENTATION

The easiest way to represent a 3D object like a box, cube, pyramid, prism, etc. is a "wireframe" representation. In this the object just consists of a number of points (called "corners" or "nodes" or "vertices", etc.) which can be joined by straight lines (called "edges" or "arcs", etc.). Suppose there are C corners and E edges, then, for a cube, C=8 and E=12.

8.1.1 A data structure

To represent an object like a cube in a computer we need to be able to store information about the corners and the edges in the form of numbers. Information about the corners will usually be sets of 3D coordinates — which means that an object must be chosen, the directions of the axes must be fixed and suitable scales need to be selected. This information is called **geometric**. The sets of coordinate will form an array inside the computer and we must be systematic in the order that the corners are picked — each corner needs to be given an "index number" between 1 and C.

Information about the edges is easier. Each edge joins just two corners, so for each edge we only need to store the index numbers of the two corners that it joins. This information is called **topological**. As with the coordinates the information on edges will be held in arrays and so each must be labelled with an index number between 1 and E.

The simplest choice for the axes is to fix the origin in the centre of the cube and to make the axes parallel to its sides. In this example the x-axis comes out towards you, the y-axis runs horizontally to the right and the z-axis runs vertically to the top of the paper. If we choose the unit on each axis to be half the cube's side then we have very simple numbers to work with:

$$C = \begin{bmatrix} 1 & 1 & 1 \\ -1 & 1 & 1 \\ -1 & -1 & 1 \\ 1 & -1 & 1 \\ 1 & 1 & -1 \\ 1 & -1 & 1 \\ -1 & -1 & -1 \\ 1 & -1 & -1 \end{bmatrix} \quad \text{start} = \begin{bmatrix} 1 \\ 2 \\ 3 \\ 4 \\ 1 \\ 2 \\ 3 \\ 4 \\ 5 \\ 6 \\ 7 \\ 8 \end{bmatrix} \quad \text{finish} = \begin{bmatrix} 2 \\ 3 \\ 4 \\ 1 \\ 5 \\ 6 \\ 7 \\ 8 \\ 6 \\ 7 \\ 8 \\ 5 \end{bmatrix}$$

Now we have now all the information needed to define the object:

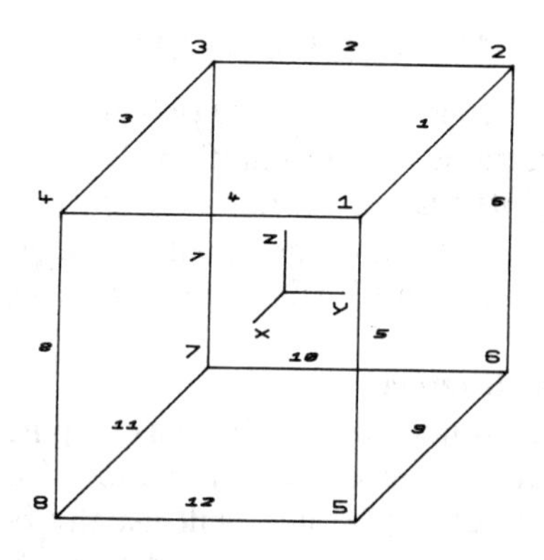

Fig. 8.1.

```
1000 DEF PROCdefine
1010  C = 8 : E = 12
1020  DIM C(C,3), start(E), finish(E)
1030  FOR corner = 1 TO C
1040   FOR j = 1 TO 3
1050    READ C(corner,j)
1060   NEXT j
1070  NEXT corner
1080  DATA 1,1,1,  -1,1,1,  -1,-1,1,  1,-1,1
1090  DATA 1,1,-1,  -1,1,-1,  -1,-1,-1,  1,-1,-1
1110  FOR edge = 1 TO E
1120   READ start(edge), finish(edge)
1130  NEXT edge
1140  DATA 1,2,  2,3,  3,4,  4,1,  1,5,  2,6
1150  DATA 3,7,  4,8,  5,6,  6,7,  7,8,  8,5
1190 ENDPROC
1195 :
```

To test that we have not made any mistakes we just need to "call" this procedure with:

```
100 PROCdefine
490 END
499 :
```

If this works without error messages than there is a fair chance that we have not made any mistakes so far. Otherwise make a careful check of the commas in the data lines. The only trouble is that computer has not yet drawn any pictures!

8.1.2 Orthographic display

To display the object we need to draw each edge on the screen. This means that for each edge we must "look-up" in the start and finish arrays the labels of the corners it joins and then "pick-up" the corresponding coordinates from the C array. We must now PROJECT these 3D coordinates into pairs of suitable 2D coordinates for the screen, and join them.

The simplest pojection is one that just ignores the depth (i.e. the x-coordinate) of each point. Then the screen x-coordinate is the space y-coordinate and the screen y-coordinate is the space z-coordinate:

```
1200 DEF PROCdisplay
1210   FOR edge = 1 TO E
1220    start = start(edge) : finish = finish(edge)
1230    Xs = C(start,2)
1240    Ys = C(start,3)
1250    X1 = Xs : Y1 = Ys
1260    Xs = C(finish,2)
1270    Ys = C(finish,3)
1280    PROCjoin(X1,Y1, Xs,Ys, FC)
1290   NEXT edge
1295 ENDPROC
1296 :
```

Just a few additions are needed to make the display:

```
10 REM Prog.8.1 - wireframe display
49 :
50 MODE 1

110 PROCdisplay
```

You might have to look very hard to see the "cube" — it is a tiny dot at the bottom left-hand corner of the screen! (or, for some micros, at the top). We need to do some scaling to make a picture of a reasonable size:

```
70 SX = 100 : SY = 100
```

We still have not moved the centre of the picture away from the bottom left-hand corner of the screen, so we need to move the origin:

```
50 MODE 1 : PROCsetup
60 OX = 0.5*SW : OY = 0.5*SH
99 :
```

where SW and SH are defined in PROCsetup as the width and height of the screen:
At last we have a picture, but it seems a lot of effort just to get a SQUARE!

Fig. 8.2 — Orthographic view of cube.

8.1.3 An oblique display

A more helpful projection is an **oblique** projection, in which points in front of the screen get swung down and to the left. This means that we need to subtract multiples of the x-coordinate from the space y- and z-coordinates when calculating the screen coordinates (Xs,Ys):

```
1180  mx = 0.5 : my = 0.5
1230   Xs = C(start,2) - mx*C(start,1)
1240   Ys = C(start,3) - my*C(start,1)
1260   Xs = C(finish,2) - mx*C(finish,1)
1270   Ys = C(finish,3) - my*C(finish,1)
```

This simple change makes a world of difference to the representation. Try

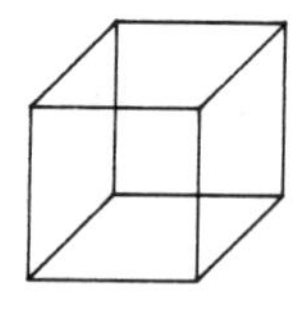

Fig. 8.3 — Oblique view of cube.

using different factors in line 1180 to vary the view.

Now we have a useable framework it should be easy to build up more complex 3D objects by extending PROCdefine. Suppose we want to put a "hat" on the cube. This will need one extra corner and four extra edges:

```
1010  C = 9 : E = 16
1095  DATA 0,0,3
1155  DATA 9,1, 9,2, 9,3, 9,4
```

Now try making up some interesting shapes of your own. Try to make a

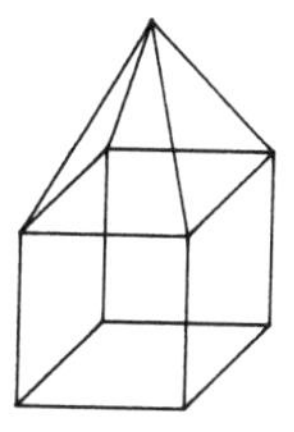

Fig. 8.4 — Cube with a roof.

reasonable "house" and then put the call to PROCdisplay into a loop that changes the origin OX,OY to get a row of houses.

8.1.4 Performing transformations

Now that the computer "knows" enough about your 3D object to represent it, it should be possible to perform transformations on it. One transformation that is simple for the computer, but very hard in real life, is a **stretch**. We just need to specify factors for how much stretch is to be applied along each of the three axes:

```
1300 DEF PROCstretch(fx,fy,fz)
1310   FOR corner = 1 TO C
1320     C(corner,1) = C(corner,1)*fx
1330     C(corner,2) = C(corner,2)*fy
1340     C(corner,3) = C(corner,3)*fz
1350   NEXT corner
1390 ENDPROC
1395 :
```

A few simple additions bring this into play:

```
120 INPUT fx, fy, fz
130 PROCstretch(fx,fy,fz)
140 PROCdisplay
```

How would you use this to get a one-way stretch, or an enlargement?

A very useful transformation is given by a **rotation**. In 3D we need to know which is the **axis** of the rotation, and through what angle to turn. Suppose we do a rotation about the z-axis, then all the z-coordinates will be unaltered and the x- and y-coordinates will be altered in the same way as for a 2D rotation:

```
1400 DEF PROCrotz(ang)
1410   arad = ang*PI/180 : cos = COS(arad) : sin = SIN(arad)
1420   FOR corner = 1 TO C
1430     x = C(corner,1) : y = C(corner,2)
```

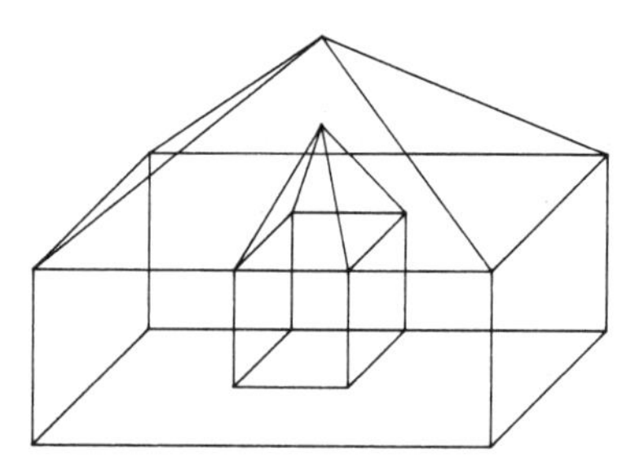

Fig. 8.5 — A stretched cube.

```
1440    C(corner,1) = x*cos - y*sin
1450    C(corner,2) = x*sin + y*cos
1460   NEXT corner
1490 ENDPROC
1495 :
```

This can easily be tested by adding:

```
150 INPUT ang
160 PROCrotz(ang)
170 PROCdisplay
```

Try to build up a set of procedures that perform interesting 3D transforma-

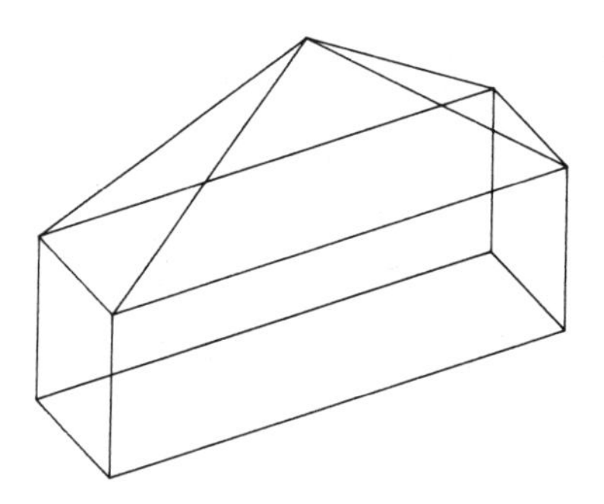

Fig. 8.6 — A rotated cube.

tions. You could define separate ones for PROCrotx and PROCroty, or you could try to make one procedure that can rotate about any given axis.

8.1.5 A perspective projection

It is quite surprisingly simple to change the projection without changing many lines of the program. We have met **orthographic**, **oblique** and **isometric** projections. For a **perspective** projection we have to decide how far a **viewpoint** is away from the screen. Suppose we imagine one of our eyes at the viewpoint V with 3D coordinates (vp,0,0). Then the screen coordinates (Xs,Ys) of a point will be the y- and z-space coordinates enlarged by a factor which depends upon how near the point is to the eye. This factor must tend to infinity as the point's x-coordinate approaches V. A suitable factor is given by: vp/(vp−x).

Once a value for vp has been defined the changes to PROCdisplay are

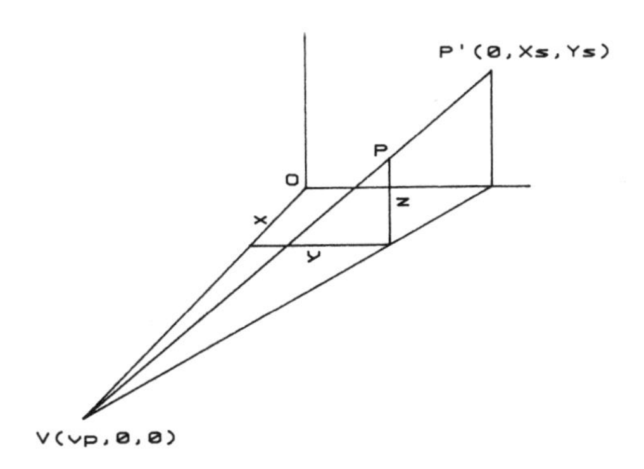

Fig. 8.7 — Diagram for perspective.

simple:

```
1180  vp = 4
1225   m  = vp/(vp-C(start,1))
1230   Xs = C(start,2)*m
1240   Ys = C(start,3)*m
1255   m  = vp/(vp-C(finish,1))
1260   Xs = C(finish,2)*m
1270   Ys = C(finish,3)*m
```

Try different choices for vp to vary the perspective. If vp is very large the image approaches orthographic projection. If vp is only a little larger than the largest x-coordinate then you get a very distorted picture. If P is "inside" the object then very weird things happen! You can prevent such effects by putting "traps" in lines 1222 and 1252 such as:

```
1185  delta = 0.1
1222   IF C(start,1) >(vp-delta) THEN GOTO 1290 : REM next edge
1252   IF C(finish,1)>(vp-delta) THEN GOTO 1290 : REM next edge
```

8.1.6 Isometric projection

As a final touch some illustration are shown in **isometric** projection. To examine this the changes to the program are again small:

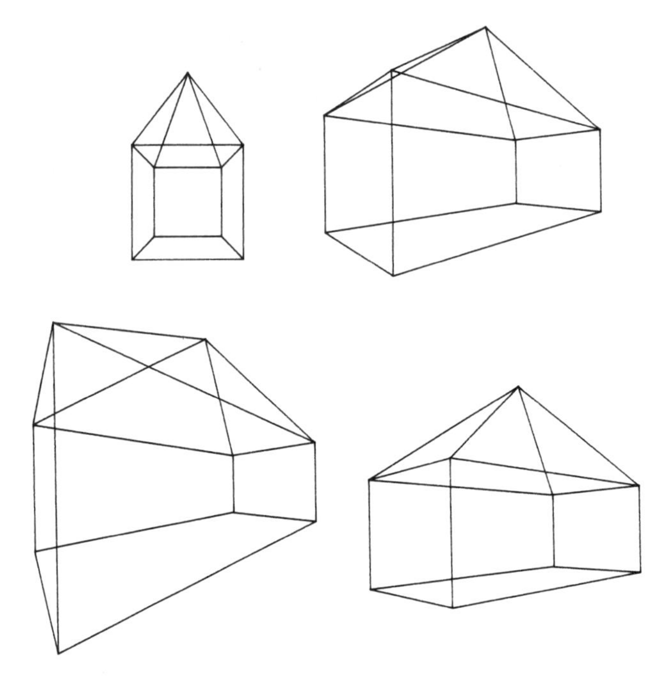

Fig. 8.8 — Perspective views of a cube.

```
1180  mx = COS(PI/6) : my = SIN(PI/6)
1185 :
1222 :
1225 :
1230   Xs = (C(start,2) - C(start,1))*mx
1240   Ys = C(start,3) - (C(start,1) + C(start,2))*my
1252 :
1255 :
1260   Xs = (C(finish,2) - C(finish,1))*mx
1270   Ys = C(finish,3) - (C(finish,1) + C(finish,2))*my
```

Now we have all the components to make up a package that could offer a choice of objects on which you could perform a choice of transformations and display the result in a choice of projections.

The major limitation imposed is that the object is represented as a **wireframe** structure and shown in just one colour. How could you represent a cylinder, cone or sphere? Can you make different edges of a structure be shown in different colours? Could you store information about the **faces** of the body that would enable you to colour them in? You would have to be careful to colour in working from the back to the front of the object or else you would get some strange results.

8.2 ADDING COLOUR AND VIEWING IN STEREO

8.2.1 Colouring the edges

The **wireframe** skeleton of the solid developed in the last section was only drawn in a single colour. It was defined by a simple data structure consisting

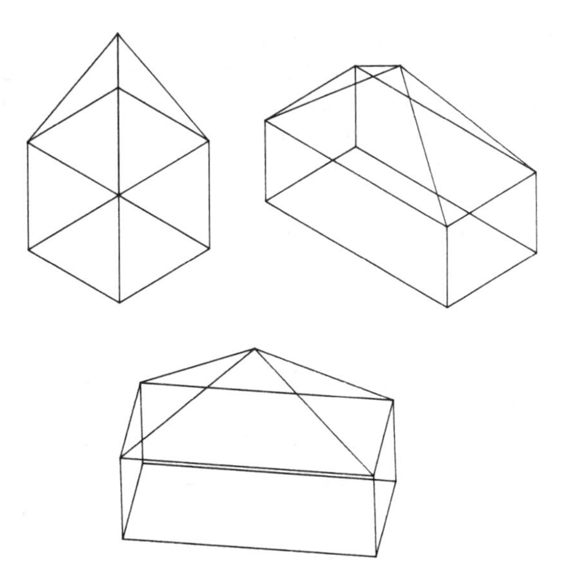

Fig. 8.9 — Some isometric projections.

of the three arrays (or matrices) C, 'start' and 'finish', containing information about the position of vertices and the incidence of vertices on edges we could easily arrange for each edge to be associated with a particular colour by extending the data structure slightly. We just need a list (or vector) of colour codes for each of the E edges. Suppose this list is called 'paint', then the changes to PROCdefine are simple:

```
1025  DIM paint(E)
1160  FOR edge = 1 TO E
1165   READ paint(edge)
1170  NEXT edge
1175  DATA 3,3,3,3,1,1,1,1,2,2,2,2
```

This would define the four edges in the top plane of the cube to be in colour 3 (white), the four vertical edges to be in colour 1 (red) and the four edges in the bottom plane be in colour 2 (yellow).

Now we have the colours defined we need to "look them up" in PROCdisplay:

```
1280     PROCjoin(X1,Y1, Xs,Ys, paint(edge))
```

and that is all that is needed to achieve a colour display. Try adding colour to some more complex shapes.

8.2.2 Colouring stereo images

An astonishing but easy application of these techniques is making a display which can be viewed through the cheap coloured cardboard and plastic 3D viewers given away in magazines. Usually the left eye looks through a red filter and the right eye through a blue or green filter. When we have displayed a solid in perspective projection we have assumed that there is a single "eye" looking at it at the viewpoint. In fact, for binocular vision, there are two viewpoints each displaced horizontally slightly. The right eye may be able to see parts of the right side of the body hidden to the left eye and vice versa. The image presented to the right eye is that of the cube in monocular vision **rotated** slightly about the z-axis.

Thus we have a simple way of displaying two stereo pictures of the object:

First define the object
then rotate it through a small angle A to the left
then display it in the colour for the right eye (blue)
then rotate it through 2*A to the right
then display it in the colour for the left eye (red)

It may be necessary to shift the origin slightly for each of the two views to avoid too much interference on the screen. PROCpalette redefines the colour whose code is 2 to be blue (rather than the default value of yellow). Since this is a piece of machine specific code it is given five-figure line numbers. For the BBC micro use VDU 19,2,2,0,0,0 if you want green rather than blue:

```
   10 REM Prog.8.2 - Stereo view, Right = Blue, Left = Red
   49 :
   50 MODE 1 : PROCsetup
   60 OX = 0.5*SW : OY = 0.5*SH
   70 SX = 100 : SY = 100
   80 PROCpalette
   99 :
  100 PROCdefine
  110 PROCrotz(-4)
  120 OX = OX + 10
  130 PROCdisplay(2)
  140 PROCrotz(8)
  150 OX = OX - 20
  160 PROCdisplay(1)
  490 END

 1200 DEF PROCdisplay(PC)
 1280     PROCjoin(X1,Y1, Xs,Ys, PC)

10100 DEF PROCpalette
10110  VDU 19,2,4,0,0,0
10190 ENDPROC
```

For the program to run you will need to include PROCdefine, PROCdisplay and PROCrotz for your chosen shape. PROCdisplay should use a perspective projection. You may need to experiment with the value of vp, the size of the small rotation angle and the shifts of origin to get a really impressive display.

8.3 COLOURING FACES

Instead of basing our data structure on representing edges we could represent the faces. There is one very big difference, though. We know that

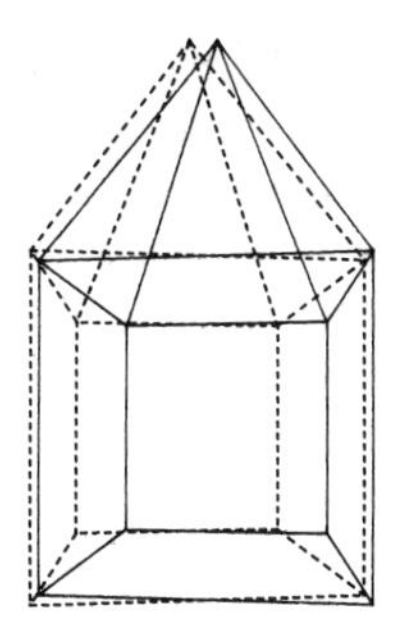

Fig. 8.10 — A stereo pair.

for all solids each edge will just connect two vertices. However, a face of a solid may be a triangle, a quadrilateral or some other polygon. We cannot assume, except in special cases, that all faces will contain the same number of vertices. A general and efficient solution to developing an appropriate data structure for this case needs techniques beyond the scope of this book. Instead we shall stick to using the simple data structure of an array but carry an overhead of some redundancy.

Suppose the body has F faces (F=6 for the cube) and that we know that no face contains more than MF vertices. For each face we need to supply a list of at least three and at most MF vertices that are incident on it. If the face contains less than MF vertices we could either "pad-out" the list with a string of 0's or just use a single 0 as a code to tell us that there are no more to come.

For the cube the back face has vertices 2, 3, 7 and 6 on its boundary. If we store the faces in the order: back, bottom, right, top, left, front then we can make up a 'bound' matrix:

$$\text{bound} = \begin{bmatrix} 2 & 3 & 7 & 6 \\ 5 & 6 & 7 & 8 \\ 5 & 6 & 2 & 1 \\ 1 & 2 & 3 & 4 \\ 3 & 4 & 8 & 7 \\ 1 & 4 & 8 & 5 \end{bmatrix}$$

The following additions illustrate the process (though remember that they are rather "over-kill" for the cube which has all its faces containing exactly four vertices!):

```
1005  PROCfaces

1500 DEF PROCfaces
1510  F = 6 : MF = 8 : DIM bound(F,MF)
1520  FOR face = 1 TO F
1530   corner = 0 :
1540   REPEAT
1550    corner = corner + 1
1560    READ bound(face,corner)
```

```
1570    UNTIL bound(face,corner)<1 OR corner=MF
1580   NEXT face
1590   DATA 2,3,7,6,0, 5,6,7,8,0, 5,6,2,1,0
1600   DATA 1,2,3,4,0, 3,4,8,7,0, 1,4,8,5,0
1610   DIM fill(F)
1620   FOR face = 1 TO F
1630    READ fill(face)
1640   NEXT face
1650   DATA 1,2,2,2,2,3
1690 ENDPROC
1695 :
```

For good measure we have also added an array called 'fill' to hold the colour codes for the F faces.

In order to display the result we must be able to colour in the polygons defined in the 'bound' array. Suppose we are colouring in the i-th face. Then bound(i,1) holds the label of the starting vertex Vs and bound(i,2) holds the label of the next vertex Vn. From the coordinate matrix C we can find the space coordinates (x,y,z) of Vs and transform these to the 2D screen coordinates (Xs, Ys). We now want to move to (Xs, Ys), compute the screen coordinate of Vn and draw an edge from Vs to Vn.

Then we can let the point Vn move round each of the remaining vertices of the face in turn, always colouring in the triangle formed by the new position of Vn, the previous position of Vn and the starting vertex Vs. In this way we can colour in the whole polygonal face. We must stop when the label of Vn is either 0 or we have visited MF vertices:

```
 110 PROCface_display

4000 DEF PROCface_display
4010  FOR face = 1 TO F
4020   PC = fill(face)
4030   Vs = bound(face,1) : Vn = bound(face,2)
4040   x = C(Vs,1) : y = C(Vs,2) : z = C(Vs,3)
4050   X1 = y - x*mx : Y1 = z - x*my
4060   x = C(Vn,1) : y = C(Vn,2) : z = C(Vn,3)
4070   X2 = y - x*mx : Y2 = z - x*my
4080   PROCjoin(X1,Y1, X2,Y2, PC)
4090   corner = 3 : Vn = bound(face,3)
4100   REPEAT
4110    x = C(Vn,1) : y = C(Vn,2) : z = C(Vn,3)
4120    X3 = y - x*mx : Y3 = z - x*my
4130    PROCfill(X1,Y1, X2,Y2, X3,Y3, PC)
4140    X2 = X3 : Y2 = Y3
4150    corner = corner + 1 : Vn = bound(face,corner)
4160   UNTIL Vn<1 OR corner=MF
4170  NEXT face
4190 ENDPROC
```

We have arranged for the faces to be "painted" in a particular order. See what happens as you rotate the cube. Try displaying the cube in other projections. Try to model shapes, such as the cube with a "hat" or triangular prisms, that have different shaped faces. Can you define a solid with one or more non-convex faces? Clearly, whichever representation we choose, it would be nice to have a means of detecting which edges or faces are hidden from the eye, and to avoid displaying them.

8.4 HIDDEN LINE REMOVAL

This is a classic problem and a general algorithm for any solid is beyond the scope of both this book and most micros. However, there is a general algorithm for **convex** bodies in perspective based upon vector geometry. First, though, the simple data structure must be extended to give topological information about the incidence of vertices and edges on faces.

The algorithm depends upon distinguishing the faces of the body as either **hidden** or **visible**. Each edge belongs to two faces so that an edge is **hidden** if and only if both its faces are **hidden**. To find whether a face is hidden we need to know the coordinates of three successive points A, B, C on its perimeter. Using these we can compute the vectors:

$$\mathbf{p} = \mathbf{BA} \qquad \text{and} \qquad \mathbf{q} = \mathbf{BC}$$

If A, B. C are in clockwise order when viewed from **outside** the face then the vector product:

$$\mathbf{n} = \mathbf{BA} \times \mathbf{BC} = \mathbf{p} \times \mathbf{q}$$

will be an **outward** pointing normal to the surface (sometimes called a 'flagpole'). If the eye is at the point E with coordinates (vp, 0, 0) then we can compute the vector $\mathbf{e} = \mathbf{BE}$ from the base of the flagpole B to the eye E. The face will be hidden if the angle between **n** and **e** is more than a right angle (i.e it points away from the viewer). A simple test for this is to form the scalar product $s = \mathbf{n.e}$ and to test the sign of s : if $s<0$ then the face is hidden.

In addition to the arrays C, 'start' and 'finish', containing information about vertices and edges we shall need two further arrays 'node' and 'face' to hold the minimum amount of information about the faces. For the cube there are 6 faces and these are labelled as shown in Fig. 8.11.

For each face we note the labels of any three vertices met in clockwise

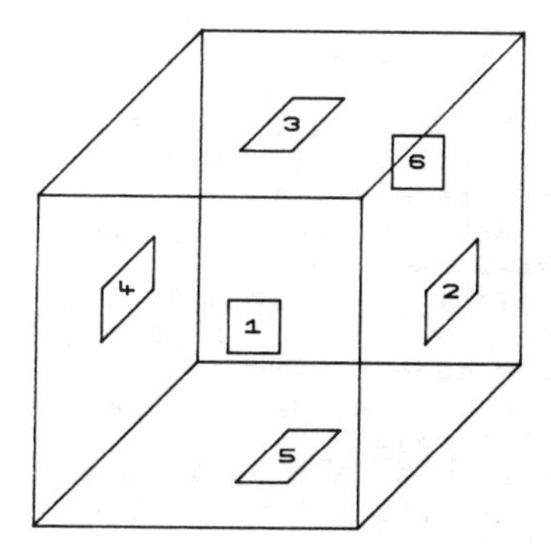

Fig. 8.11 — Faces of a cube.

order when viewed from outside the body e.g.:

$$\text{node} = \begin{bmatrix} 1 & 5 & 8 \\ 1 & 2 & 6 \\ 1 & 4 & 3 \\ 3 & 4 & 8 \\ 8 & 5 & 6 \\ 6 & 2 & 3 \end{bmatrix}$$

The order in which the faces are taken is unimportant, this time. For each edge of the body we note the labels of the two faces it separates:

$$\text{face} = \begin{bmatrix} 2 & 3 \\ 3 & 6 \\ 3 & 4 \\ 1 & 3 \\ 1 & 2 \\ 2 & 6 \\ 4 & 6 \\ 1 & 4 \\ 2 & 5 \\ 5 & 6 \\ 4 & 5 \\ 1 & 5 \end{bmatrix}$$

As in the last section we shall store the number of faces in F. The changes in the data structure section are, then:

```
 105 PROCfaceinfo

1500 DEF PROCfaceinfo
1510   F = 6
1520   DIM node(F,3), face(E,2)
1530   DIM p(3), q(3), n(3), e(3)
1540   FOR face = 1 TO F
1550    READ node(face,1), node(face,2), node(face,3)
1560   NEXT face
1570   DATA 1,5,8, 1,2,6, 1,4,3
1580   DATA 3,4,8, 8,5,6, 6,2,3
1590   FOR edge = 1 TO E
1600    READ face(edge,1), face(edge,2)
1610   NEXT edge
1620   DATA 2,3, 3,6, 3,4, 1,3, 1,2, 2,6
1630   DATA 4,6, 1,4, 2,5, 5,6, 4,5, 1,5
1690 ENDPROC
1695 :
```

The vectors **p**, **q**, **n**, and **e** will be used in deciding whether a face is hidden or not.

The algorithm to mark the faces as **hidden** or **visible** becomes:

```
1700 DEF PROCcheck
1710  FOR face = 1 TO F
1720   node(face,0) = 1
1730   AA = node(face,1) : BB = node(face,2) : CC = node(face,3)
1740   FOR j = 1 TO 3
1750    p(j) = C(AA,j) - C(BB,j)
1760    q(j) = C(CC,j) - C(BB,j)
1770   NEXT j
1780   n(1) = p(2)*q(3) - p(3)*q(2)
1790   n(2) = p(3)*q(1) - p(1)*q(3)
1800   n(3) = p(1)*q(2) - p(2)*q(1)
1810   e(1) = vp - C(BB,1) : e(2) = -C(BB,2) : e(3) = -C(BB,3)
1820   s = e(1)*n(1) + e(2)*n(2) + e(3)*n(3)
1830   IF s<0 THEN node(face,0) = 0
1840  NEXT face
1890 ENDPROC
1895 :
```

Here we use the fact that most microcomputer versions of Basic use index numbers starting from zero when you dimension an array using DIM. Thus there is a complete "first column" of unused space in 'node': node(1,0), node(2,0), node(3,0), . . . which we can use to store the code which is used to denote whether the i-th face is hidden (code 0) or visible (code 1). We could have used another column of 'node' for the purpose or we could have used a separate list or vector.

In order to avoid drawing hidden lines we just need to add the simple test:

```
1205  PROCcheck
1214   v1 = node(face(edge,1),0)
1215   v2 = node(face(edge,2),0)
1216   IF (v1=0) AND (v2=0) THEN GOTO 1290 : REM next edge
```

The i-th edge belongs to the two faces face(i,1) and face(i,2) and will only be hidden if both those faces are marked as hidden. The projection used in PROCdisplay must be **perspective** if the results are to look sensible.

Try using these ideas together with perspective projection to represent different **convex** bodies, e.g. put a roof on the cube. Try it with transformations such as rotations. Beware that certain transformations, such as reflections, might turn a clockwise order into an anticlockwise one — what adjustments would be needed? Can you arrange for the hidden lines to be shown as dotted lines? Can you arrange just to colour in the visible faces.

8.5 REGULAR SOLIDS

As we started with a cube it might be useful to have definitions for the other four regular (Platonic) solids. The next most obvious one is the octahedron, which is the dual of the cube.

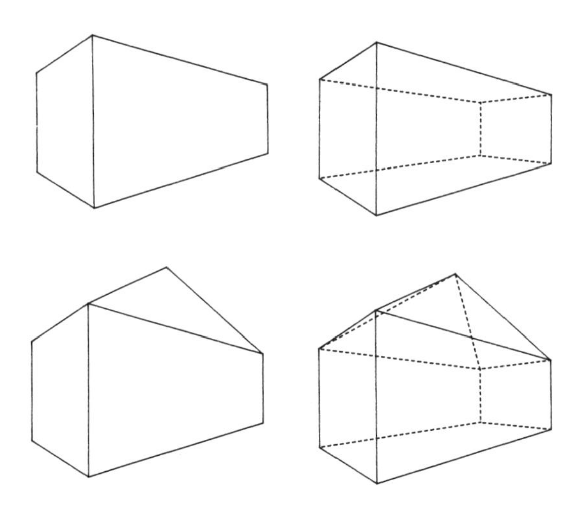

Fig. 8.12 — Some displays with hidden lines removed and dotted.

8.5.1 The octahedron

The octahedron has six vertices which we can site as the centres of the six faces of the cube. The front face of the cube was taken as the plane x=1 and so its midpoint is (1, 0, 0). Similarly, we can find coordinates for the other five vertices. We need to join point 1 to each of the points 2, 3, 4 and 5. They need to be joined in a "ring": 2–3, 3–4, 4–5, 5–1. Finally, each of the points 2, 3, 4, 5 needs to be joined to point 6. The only changes needed for a wireframe display are in PROCdefine which we give in full:

```
1000 DEF PROCdefine
1010  C = 6 : E = 12
1020  DIM C(C,3), start(E), finish(E)
1030  FOR corner = 1 TO C
1040   FOR j = 1 TO 3
1050    READ C(corner,j)
1060   NEXT j
1070  NEXT corner
1080  DATA 1,0,0, 0,1,0, 0,0,1, 0,-1,0
1090  DATA 0,0,-1, -1,0,0
1110  FOR edge = 1 TO E
1120   READ start(edge), finish(edge)
1130  NEXT edge
1140  DATA 1,2, 1,3, 1,4, 1,5, 2,3, 3,4
1150  DATA 4,5, 5,2, 2,6, 3,6, 4,6, 5,6
1190 ENDPROC
```

In order to remove hidden lines we must also define the face data for the arrays 'node' and 'face'. The octahedron has eight triangular faces:

```
1510  F = 8

1570  DATA 1,3,2, 2,3,6, 6,3,4, 4,3,1
1580  DATA 1,2,5, 2,6,5, 6,4,5, 4,1,5

1620  DATA 1,5, 1,4, 4,8, 5,8, 1,2, 4,3
1630  DATA 7,8, 5,6, 6,2, 2,3, 3,7, 6,7
```

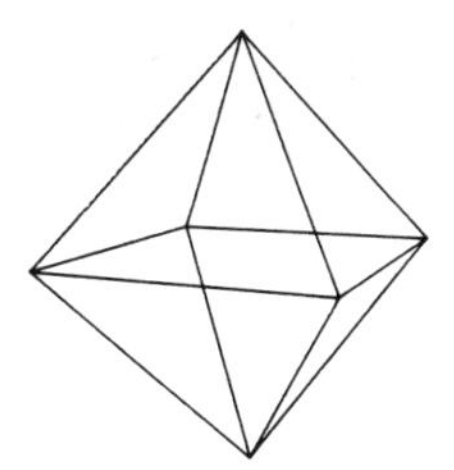

Fig. 8.13 — Wire frame octahedron.

See if you can adapt the display to colour in the faces (you will need to

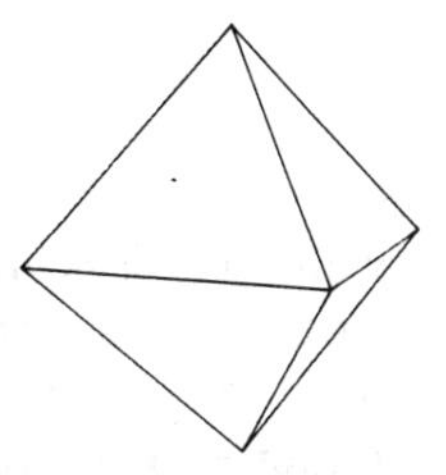

Fig. 8.14 — Octahedron with hidden lines removed.

redefine the data in PROCfaces).

8.5.2 The tetrahedron

The tetrahedron has only four vertices, six edges and four triangular faces. The data for the C, 'start' and 'finish' matrices can be defined as:

```
1010  C = 4 : E = 6

1080  DATA 1,1,1, -1,-1,1
1090  DATA 1,-1,-1, -1,1,-1

1140  DATA 1,2, 2,3, 3,1
1150  DATA 1,4, 4,3, 4,2
```

and the corresponding data for the face matrices 'node' and 'edge' is:

```
1510  F = 4

1570  DATA 1,3,2, 1,4,3
1580  DATA 1,2,4, 2,3,4

1620  DATA 1,3, 1,4, 1,2
1630  DATA 2,3, 2,4, 3,4
```

This gives rather a strange view of a tetrahedron and you might prefer either to perform some initial transformations on C or to redefine the coordinates.

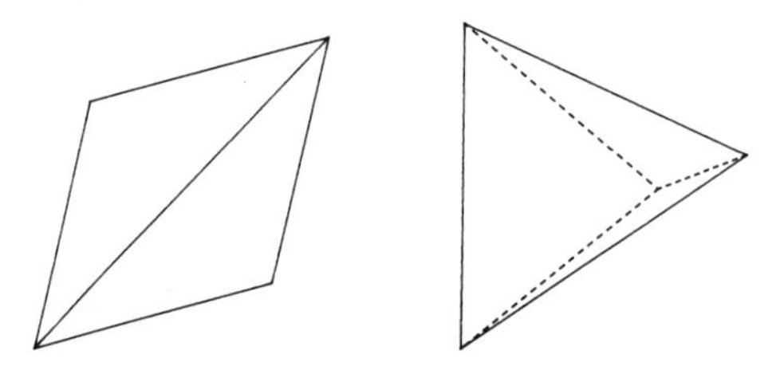

Fig. 8.15 — Views of a tetrahedron.

8.5.3 The icosahedron

The icosahedron has twelve vertices, thirty edges and twenty triangular faces. The coordinates can be given a pattern by using a simple trick. If we let **ro** be the **golden ratio, ro** = (1+SQR(5))/2, then each point will have one of its coordinates equal to either +**ro** or −**ro**; one equal to either +1 or −1 and the other equal to 0. Rather than repeatedly typing 1.618033988 in the DATA statements we can use the idea of "codes". Wherever we want ro to be used we can just insert a code (such as 2) and decode it in PROCdefine:

```
1010  C = 12 : E = 30
1015  ro = (1+SQR(5))/2

1050   READ da
1051   IF ABS(da)=2 THEN da = SGN(da)*ro
1052   C(corner,j) = da

1080  DATA 1,0,2, 0,2,1, -1,0,2, 0,-2,1, 2,1,0, 2,-1,0
1090  DATA 1,0,-2, 0,2,-1, -2,1,0, -1,0,-2, -2,-1,0, 0,-2,-1

1140  DATA 1,2, 2,3, 3,1, 1,5, 5,8, 8,2, 2,5, 8,9
1145  DATA 9,2, 9,3, 4,1, 1,6, 6,5, 5,7, 7,8, 8,10
1150  DATA 10,9, 9,11, 11,3, 3,4, 4,11, 11,10, 10,7, 7,6
1155  DATA  6,4, 4,12, 12,11, 10,12, 12,7, 6,12

1510  F = 20

1570  DATA 2,1,3, 2,5,1, 8,5,2, 8,2,9, 2,3,9, 1,4,3, 1,6,4, 5,6,1
1575  DATA 5,7,6, 8,7,5, 8,10,7, 9,10,8, 11,10,9, 3,11,9, 4,11,3
1580  DATA 4,12,11, 12,4,6, 12,6,7, 10,12,7, 11,12,10

1620  DATA 1,2, 1,5, 1,6, 2,8, 3,10, 3,4, 2,3, 4,12
1625  DATA 4,5, 5,14, 6,7, 7,8, 8,9, 9,10, 10,11, 11,12
1630  DATA 12,13, 13,14, 14,15, 15,6, 15,16, 13,20, 11,19
1635  DATA 9,18, 7,17, 16,17, 16,20, 19,20, 18,19, 17,18
```

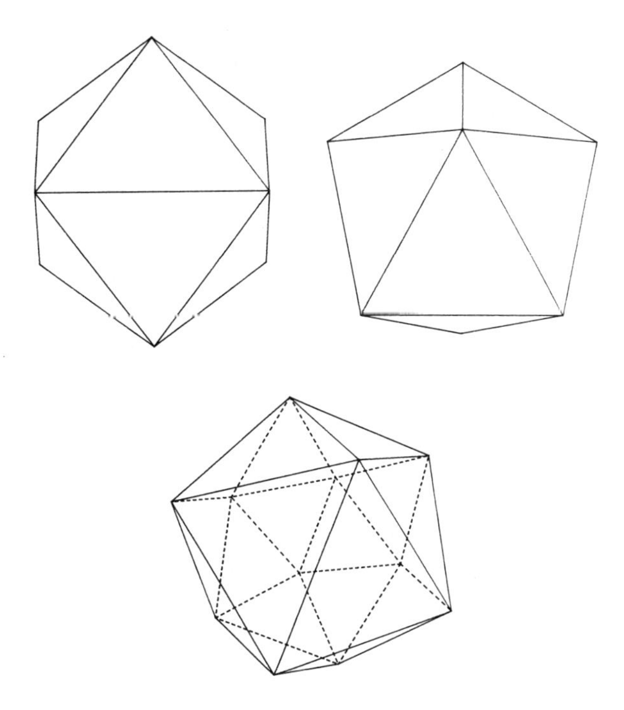

Fig. 8.16 — Views of icosahedra.

8.5.4 The dodecahedron

The dodecahedron has twenty vertices, thirty edges and twelve faces. We can use the same technique as for the icosahedron to "code" the data. In this case, though, some of the vertices have coordinates equal to 1/**ro** rather than **ro**:

```
1010  C = 20 : E = 30
1015  ro = (1+SQR(5))/2

1050   READ da : sg = SGN(da) : ab = ABS(da)
1051   IF ab=2 THEN da = sg*ro
1052   IF ab=3 THEN da = sg/ro
1053   C(corner,j) = da

1080  DATA 2,0,3, 2,0,-3, 1,-1,-1, 3,-2,0, 1,-1,1, 1,1,1
1085  DATA 3,2,0, 1,1,-1, 0,3,-2, 0,-3,-2, -1,-1,-1, -3,-2,0
1090  DATA -1,-1,1, 0,-3,2, 0,3,2, -1,1,1, -3,2,0, -1,1,-1
1095  DATA -2,0,-3, -2,0,3

1140  DATA 1,2, 2,3, 3,4, 4,5, 5,1, 1,6, 2,8, 3,10
1145  DATA 4,12, 5,14, 14,15, 15,6, 6,7, 7,8, 8,9, 9,10
1150  DATA 10,11, 11,12, 12,13, 13,14, 15,16, 7,17, 9,18, 11,19
1155  DATA 13,20, 20,16, 16,17, 17,18, 18,19, 19,20

1510  F = 12

1570  DATA 1,2,3, 10,11,12, 12,13,14, 14,15,6
1575  DATA 6,7,8, 8,9,10, 19,20,13, 15,14,13
1580  DATA 15,16,17, 17,18,9, 9,18,19, 17,16,20
```

```
1620  DATA 1,5, 1,6, 1,2, 1,3, 1,4, 4,5, 5,6, 2,6
1625  DATA 2,3, 3,4, 4,8, 4,9, 5,9, 5,10, 6,10, 6,11
1630  DATA 2,11, 2,7, 3,7, 3,8, 8,9, 9,10, 10,11, 7,11
1635  DATA 8,7, 8,12, 9,12, 10,12, 11,12, 7,12
```

Try representing some of the simpler variations such as the truncated (or "snub") solids. If you are really ambitious you might try the stellations.

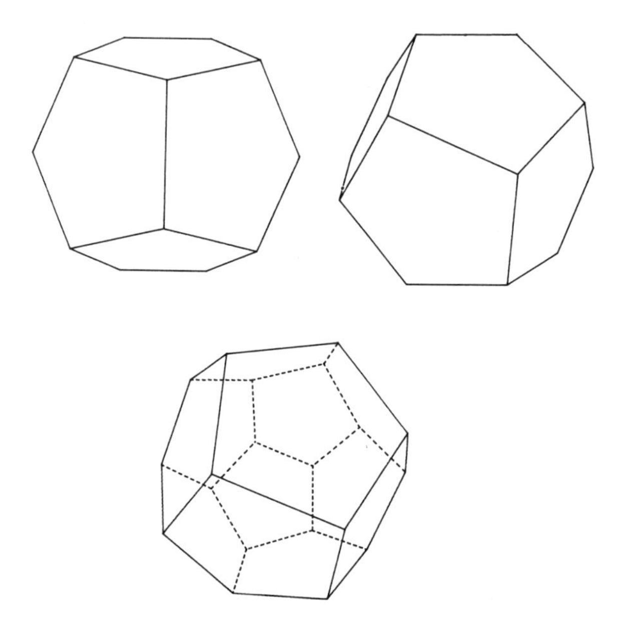

Fig. 8.17 — Some views of dodecahedra.

8.6 THE FOURTH DIMENSION (AND BEYOND)!

The techniques that we have used to represent 3D objects can be generalised to more dimensions by simply increasing the number of columns in some of the matrices used. For example, the matrix C which holds information about the coordinates of vertices could easily be dimensioned by: DIM C(C,4) to hold 4D coordinates. If an 'edge' in 4D is still to be a one-dimensional object specified by two points then the edge matrices 'start' and 'finish' do not need to be extended. Thus, for example, we can quite simply extend our structure to represent a four-dimensional object such as a **hypercube**. This has sixteen vertices consist of all possible combinations of 1's and −1's taken four at a time:

```
 10 REM Prog 8.3 - Four Dimensional Representation
 49 :
 50 MODE 1 : PROCsetup
 60 OX = 0.5*SW : OY = 0.5*SH
 70 SX = 100 : SY = 100
 99 :
100 PROCdefine
490 END
```

```
 499 :
1500 DEF PROCdefine
1510  C = 16 : E = 32
1520  DIM C(C,4), start(E), finish(E)
1530  FOR corner = 1 TO C
1540   FOR entry = 1 TO 4
1550     READ C(corner,entry)
1560   NEXT entry
1570  NEXT corner
1580  DATA 1,1,1,1, 1,-1,1,1, -1,-1,1,1, -1,1,1,1
1581  DATA 1,1,-1,1, 1,-1,-1,1, -1,-1,-1,1, -1,1,-1,1
1582  DATA 1,1,1,-1, 1,-1,1,-1, -1,-1,1,-1, -1,1,1,-1
1583  DATA 1,1,-1,-1, 1,-1,-1,-1, -1,-1,-1,-1, -1,1,-1,-1
1590  FOR edge = 1 TO E
1600   READ start(edge), finish(edge)
1610  NEXT edge
1620  DATA 1,9, 2,10, 3,11, 4,12, 5,13, 6,14, 7,15, 8,16
1621  DATA 1,5, 2,6, 3,7, 4,8, 9,13, 10,14, 11,15, 12,16
1622  DATA 1,2, 3,4, 5,6, 7,8, 9,10, 11,12, 13,14, 15,16
1623  DATA 1,4, 2,3, 5,8, 6,7, 9,12, 10,11, 13,16, 14,15
1690 ENDPROC
1699 :
```

The next task is to come up with a sensible way of mapping a point (x, y, z, w) in 4-space onto a point (Xs,Ys) in the two-dimensional space of the screen. A very simple way should be a 'double orthographic' projection in which we just ignore two of the four coordinates of each point. However, as we saw for the simple projection of a cube, this leaves a lot to be desired. Another tack is to generalise the idea that we used for oblique projection. In this case we can arrange for any two of the coordinates to be modified by adding or subtracting multiples of the other two coordinates. Thus we could have a mapping such as:

$$Xs = x \qquad + a.z + b.w$$

$$Ys = \qquad y + c.z + d.w$$

Then we can scale and shift the values of X and Y by suitable amounts for our display screen, using SX,SY, OX,OY:

```
 110 PROCdisplay

1650  m1 = 0.8 : m2 = 0.3 : n1 = 0.5 : n2 = 0.5

1700 DEF PROCdisplay
1701  REM Double Oblique Projection from 4D to 2D
1710  FOR edge = 1 TO E
1720   start = start(edge) : finish = finish(edge)
1730   Xs = C(start,1) - m1*C(start,3) - n1*C(start,4)
1740   Ys = C(start,2) - m2*C(start,3) - n2*C(start,4)
1750   X1 = Xs : Y1 = Ys
1760   Xs = C(finish,1) - m1*C(finish,3) - n1*C(finish,4)
1770   Ys = C(finish,2) - m2*C(finish,3) - n2*C(finish,4)
1780   PROCjoin(X1,Y1, Xs,Ys, FC)
1790   X1 = Xs : Y1 = Ys
1800  NEXT edge
1890 ENDPROC
1899 :
```

In order to perform transformations on our 4D object we can try to

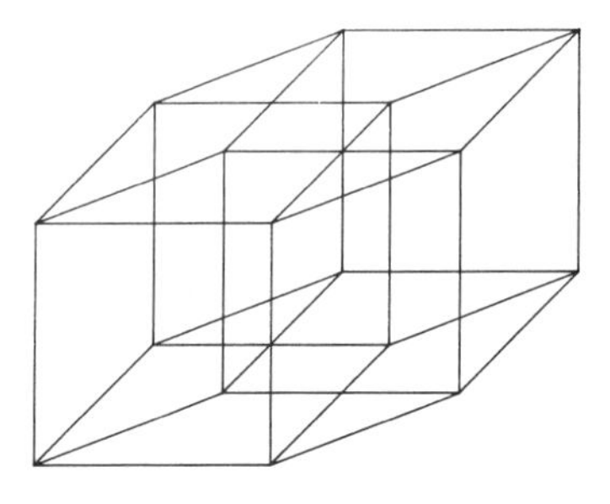

Fig. 8.18 — View of a hypercube.

generalise the transformations with which we are already familiar. For example it might well be that a 4D generalisation of a rotation can be represented by a matrix such as:

$$\begin{bmatrix} 1 & 0 & 0 & 0 \\ 0 & 1 & 0 & 0 \\ 0 & 0 & \cos A & -\sin A \\ 0 & 0 & \sin A & \cos A \end{bmatrix}$$

To experiment with this we can define a general "rotation" procedure such as:

```
2000 DEF PROCrotate(a1,a2,ang)
2010   cr = COS(ang*PI/180) : sr = SIN(ang*PI/180)
2020   FOR corner = 1 TO C
2030    C1 = C(corner,a1) : C2 = C(corner,a2)
2040    C(corner,a1) = C1*cr - C2*sr
2050    C(corner,a2) = C1*sr + C2*cr
2060   NEXT corner
2090 ENDPROC
```

and we can set the body spinning with an addition such as:

```
120 REPEAT
130   PROCrotate(3,4,10)
140   CLS
150   PROCdisplay
160   FOR Delay = 1 TO 500
170    :
180   NEXT Delay
190 UNTIL "Cows" = "home"
```

This would seem to imply that a rotation is about a **plane** rather than an axis

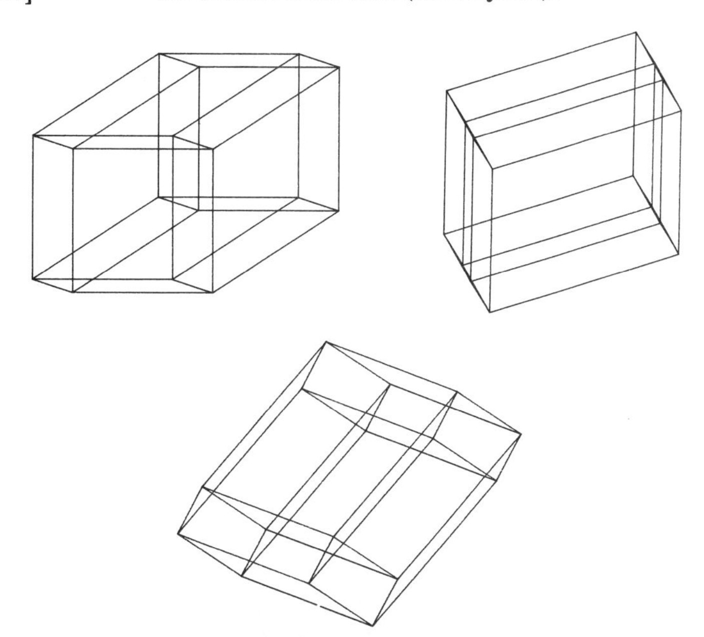

Fig. 8.19 — Some rotated hypercubes.

— does this seem sensible? Try other transformations — reflections, translations, stretches, shears!

Make the projection subroutine more general so that any choice of two coordinates can be 'lost'. Is it possible to generalise other systems of projection such as perspective or isometric? Do the ideas of "hidden lines" generalise?

Can you similarly generalise other solids like the hyper-octahedron etc.?

What is there to stop the idea being extended to more dimensions still?

9

Postscript

In this short book I have only been able to convey a few of the geometric activities that are accessible with a micro.

Some activities I have had to omit because they really require facilities which are available on only a few of the common micros. For instance, **animation** requires the ability to apply colour quickly having regard to the colour that is already there on the screen. On most micros the computational speed of Basic, and the limited range of graphic techniques makes animation impracticable. However, this is possible with the **logical** colouring functions provided by the BBC Basic GCOL command, and the colour switching provided by the VDU 19 command given some pretty efficient numerical techniques. Thus it is now possible on some micros to achieve results comparable with some of the remarkable geometric films produced a few years ago. Similarly, some remarkable curves (Snowflake, Dragon, Hilbert, fractals, etc.), can be explored if the programming language allows **recursion**. This is common in Pascal and Logo, but only a few versions of Basic, such as BBC and RM Nimbus, support it. Apart from small references in the surface display and solid manipulation programs I have avoided applications requiring the filling in of areas with colour. If this is available then there are some spendid geometric posters around now that you can not merely reproduce, but use as a basis for inventing variations. With some efficient matrix manipulation routines, such as are now available in ROM for some micros, many of the geometric applications of linear algebra become readily accessible.

Some activities I have had to omit, reluctantly, for reasons of space. There is a lot of interesting mathematics to explore in the representation of **implicitly** defined functions involving two or three variables. The micro is a superb tool for exploring the **complex** plane — the techniques extend easily to geometry on the Argand diagram, to displaying plots of complex functions and to considering mappings of the complex plane. Many magazines recently have carried superb pictures of parts of a region of the complex plane known as the **Mandelbrot** set that can be obtained (with patience!) on a micro. The techniques of surface display can be used to display **catastrophe** surfaces and to show families of curves embedded in surfaces, such as illustrating the two less obvious sets of circles that Coxeter reports as lying in a torus, or displaying the conics as curves on the surface of a cone. The use of **rational quadratics** for defining the conics can be easily explored, as can the generation of **families** of conics. A fascinating challenge is to generate five random points and to draw the conic that passes through them!

Some activities have been omitted either because I have only just started to try out some ideas, as with **differential geometry**, others because I have yet to find time to engage with them, as with **projective geometry**. I shall have no difficulty whatever in occupying all the wet weekends at my disposal!

As the design of micros, and their display devices, develops we may expect to find programming languages which are more flexible and powerful, display techniques which are more sophisticated and which offer higher resolution and a greater range of colours, and hardware which has both more memory and much greater computational speed. None the less, even within the restrictions of "old-technology" micros, such as Electron, Atari and Spectrum, currently obtainable for less than £100, there is an enormous richness in the range of geometric exploration open to any of us.

Bibliography

Barnhill, R. E. and Riesenfeld, R. F., *Computer Aided Geometric Design*, Academic Press, 1975.

Bezier, P., *Numerical Control — Mathematics and Applications*, Wiley, 1972.

Chasen, S. H., *Geometric Principles and Procedures for Computer Graphic Applications*, Prentice-Hall, 1978.

Cownie, J., *Creative Graphics on the BBC Microcomputer*, Acornsoft, 1982.

Coxeter, H. S. M., *Introduction to Geometry*, Wiley, 1967.

Cundy, H. M. and Rollett, A. P., *Mathematical Models*, OUP, 1961.

Durell, C. V., *Elementary Geometry*, G. Bell, 1925.

Faux, I. D. and Pratt, M. J., *Computational Geometry for Design and Manufacture*, Ellis Horwood, 1981.

Fletcher, T. J., *Linear Algebra Through its Applications*, Van Nostrand Reinhold, 1972.

Foley, J. D. and Van Dam, A., *Fundamentals of Interactive Computer Graphics*, Addison-Wesley, 1982.

Gasson, P. C., *Geometry of Spatial Forms*, Ellis Horwood, 1983.

Lockwood, E. H., *A Book of Curves*, CUP, 1963.

Lord, E. A. and Wilson, C. B., *The Mathematical Description of Shape and Form*, Ellis Horwood, 1984.

Newman, W. M. and Sproull, R. F., *Principles of Interactive Computer Graphics*, McGraw-Hill, 1979.

Nobbs, C. G., *Elementary Calculus and Coordinate Geometry*, OUP, 1949.

Oldknow, A. J., *Graphics with Microcomputers*, Nelson, 1985.

Oldknow, A. J. and Smith, D. V., *Learning Mathematics with Micros*, Ellis Horwood, 1983.

Rogers, D. F. and Adams, J. A., *Mathematical Elements for Computer Graphics*, McGraw-Hill, 1976.

Tuckey, C. O. and Armistead, W., *Coordinate Geometry*, *Longmans*, 1962.

Weatherburn, C. E., *Elementary Vector Analysis*, Bell, 1963.

Willson, W. W., *The Mathematics Curriculum*: *Geometry*, Blackie, 1977.

Index

Mathematics and its Applications

Series Editor: G. M. BELL, Professor of Mathematics, King's College (KQC), University of London

Artmann, B. **The Concept of Number***
Balcerzyk, S. & Joszefiak, T. **Commutative Rings***
Balcerzyk, S. & Joszefiak, T. **Noetherian and Krull Rings***
Baldock, G.R. & Bridgeman, T. **Mathematical Theory of Wave Motion**
Ball, M.A. **Mathematics in the Social and Life Sciences: Theories, Models and Methods**
de Barra, G. **Measure Theory and Integration**
Bell, G.M. and Lavis, D.A. **Co-operative Phenomena in Lattice Models Vols. I & II***
Berkshire, F.H. **Mountain and Lee Waves**
Berry, J.S., Burghes, D.N., Huntley, I.D., James, D.J.G. & Moscardini, A.O. **Teaching and Applying Mathematical Modelling**
Burghes, D.N. & Borrie, M. **Modelling with Differential Equations**
Burghes, D.N. & Downs, A.M. **Modern Introduction to Classical Mechanics and Control**
Burghes, D.N. & Graham, A. **Introduction to Control Theory, including Optimal Control**
Burghes, D.N., Huntley, I. & McDonald, J. **Applying Mathematics**
Burghes, D.N. & Wood, A.D. **Mathematical Models in the Social, Management and Life Sciences**
Butkovskiy, A.G. **Green's Functions and Transfer Functions Handbook**
Butkovskiy, A.G. **Structural Theory of Distributed Systems**
Cao, Z-Q., Kim, K.H. & Roush, F.W. **Incline Algebra and Applications**
Chorlton, F. **Textbook of Dynamics, 2nd Edition**
Chorlton, F. **Vector and Tensor Methods**
Crapper, G.D. **Introduction to Water Waves**
Cross, M. & Moscardini, A.O. **Learning the Art of Mathematical Modelling**
Cullen, M.R. **Linear Models in Biology**
Dunning-Davies, J. **Mathematical Methods for Mathematicians, Physical Scientists and Engineers**
Eason, G., Coles, C.W. & Gettinby, G. **Mathematics and Statistics for the Bio-sciences**
Exton, H. **Handbook of Hypergeometric Integrals**
Exton, H. **Multiple Hypergeometric Functions and Applications**
Exton, H. ***q*-Hypergeometric Functions and Applications**
Faux, I.D. & Pratt, M.J. **Computational Geometry for Design and Manufacture**
Firby, P.A. & Gardiner, C.F. **Surface Topology**
Gardiner, C.F. **Modern Algebra**
Gardiner, C.F. **Algebraic Structures: with Applications**
Gasson, P.C. **Geometry of Spatial Forms**
Goodbody, A.M. **Cartesian Tensors**
Goult, R.J. **Applied Linear Algebra**
Graham, A. **Kronecker Products and Matrix Calculus: with Applications**
Graham, A. **Matrix Theory and Applications for Engineers and Mathematicians**
Griffel, D.H. **Applied Functional Analysis**
Griffel, D.H. **Linear Algebra***
Hanyga, A. **Mathematical Theory of Non-linear Elasticity**
Harris, D.J. **Mathematics for Business, Management and Economics**
Hoksins, R.F. **Generalised Functions**
Hoskins, R.F. **Standard and Non-standard Analysis***
Hunter, S.C. **Mechanics of Continuous Media, 2nd (Revised) Edition**
Huntley, I. & Johnson, R.M. **Linear and Nonlinear Differential Equations**
Jaswon, M.A. & Rose, M.A. **Crystal Symmetry: The Theory of Colour Crystallography**
Johnson, R.M. **Theory and Applications of Linear Differential and Difference Equations**
Kim, K.H. & Roush, F.W. **Applied Abstract Algebra**
Kosinski, W. **Field Singularities and Wave Analysis in Continuum Mechanics**
Krishnamurthy, V. **Combinatorics: Theory and Applications**
Lindfield, G. & Penny, J.E.T. **Microcomputers in Numerical Analysis**
Lord, E.A. & Wilson, C.B. **The Mathematical Description of Shape and Form**
Marichev, O.I. **Integral Transforms of Higher Transcendental Functions**
Massey, B.S. **Measures in Science and Engineering**
Meek, B.L. & Fairthorne, S. **Using Computers**
Mikolas, M. **Real Function and Orchogonal Series**
Moore, R. **Computational Functional Analysis**
Müller-Pfeiffer, E. **Spectral Theory of Ordinary Differential Operators**
Murphy, J.A. & McShane, B. **Computation in Numerical Analysis***
Nonweiller, T.R.F. **Computational Mathematics: An Introduction to Numerical Approximation**
Ogden, R.W. **Non-linear Elastic Deformations**
Oldknow, A. & Smith, D. **Learning Mathematics with Micros**
O'Neill, M.E. & Chorlton, F. **Ideal and Incompressible Fluid Dynamics**
O'Neill, M.E. & Chorlton, F. **Viscous and Compressible Fluid Dynamics***

Page, S. G. **Mathematics: A Second Start**
Rankin, R.A. **Modular Forms**
Ratschek, H. & Rokne, J. **Computer Methods for the Range of Functions**
Scorer, R.S. **Environmental Aerodynamics**
Smith, D.K. **Network Optimisation Practice: A Computational Guide**
Srivastava, H.M. & Karlsson, P.W. **Multiple Gaussian Hypergeometric Series**
Srivastava, H.M. & Manocha, H.L. **A Treatise on Generating Functions**
Shivamoggi, B.K. **Stability of Parallel Gas Flows***
Stirling, D.S.G. **Mathematical Analysis***
Sweet, M.V. **Algebra, Geometry and Trigonometry in Science, Engineering and Mathematics**
Temperley, H.N.V. & Trevena, D.H. **Liquids and Their Properties**
Temperley, H.N.V. **Graph Theory and Applications**
Thom, R. **Mathematical Models of Morphogenesis**
Toth, G. **Harmonic and Minimal Maps**
Townend, M. S. **Mathematics in Sport**
Twizell, E.H. **Computational Methods for Partial Differential Equations**
Wheeler, R.F. **Rethinking Mathematical Concepts**
Willmore, T.J. **Total Curvature in Riemannian Geometry**
Willmore, T.J. & Hitchin, N. **Global Riemannian Geometry**
Wojtynski, W. **Lie Groups and Lie Algebras***

Statistics and Operational Research

Editor: B. W. CONOLLY, Professor of Operational Research, Queen Mary College, University of London

Beaumont, G.P. **Introductory Applied Probability**
Beaumont, G.P. **Probability and Random Variables***
Conolly, B.W. **Techniques in Operational Research: Vol. 1, Queueing Systems***
Conolly, B.W. **Techniques in Operational Research: Vol. 2, Models, Search, Randomization**
Conolly, B.W. **Lecture Notes in Queueing Systems**
French, S. **Sequencing and Scheduling: Mathematics of the Job Shop**
French, S. **Decision Theory: An Introduction to the Mathematics of Rationality**
Griffiths, P. & Hill, I.D. **Applied Statistics Algorithms**
Hartley, R. **Linear and Non-linear Programming**
Jolliffe, F.R. **Survey Design and Analysis**
Jones, A.J. **Game Theory**
Kemp, K.W. **Dice, Data and Decisions: Introductory Statistics**
Oliveira-Pinto, F. **Simulation Concepts in Mathematical Modelling***
Oliveira-Pinto, F. & Conolly, B.W. **Applicable Mathematics of Non-physical Phenomena**
Schendel, U. **Introduction to Numerical Methods for Parallel Computers**
Stoodley, K.D.C. **Applied and Computational Statistics: A First Course**
Stoodley, K.D.C., Lewis, T. & Stainton, C.L.S. **Applied Statistical Techniques**
Thomas, L.C. **Games, Theory and Applications**
Whitehead, J.R. **The Design and Analysis of Sequential Clinical Trials**

**In preparation*